Angular momentum transport and pattern formation in medium- and
wide-gap turbulent Taylor-Couette flow:
An experimental study

Angular momentum transport and pattern formation in medium- and wide-gap turbulent Taylor-Couette flow: An experimental study

Von der Fakultät für Maschinenbau, Elektro- und Energiesysteme
der Brandenburgischen Technischen Universität Cottbus – Senftenberg
zur Erlangung des akademischen Grades eines
Doktors der Ingenieurwissenschaften (Dr.-Ing.)

genehmigte Dissertation

vorgelegt von

M.Sc. Andreas Froitzheim

geboren am 17. März 1988 in Wiesbaden-Dotzheim

Vorsitzender: Prof. Dr.-Ing. habil. Hon. Prof. (NUST) Dieter Bestle
Gutachter: Prof. Dr.-Ing. Christoph Egbers
Gutachter: Prof. Dr. rer. nat. habil. Detlef Lohse
Tag der mündlichen Prüfung: 28.02.2019

Bibliografische Information der Deutschen Nationalbibliothek
Die Deutsche Nationalbibliothek verzeichnet diese Publikation in der
Deutschen Nationalbibliografie; detaillierte bibliographische Daten
sind im Internet über http://dnb.d-nb.de abrufbar.
1. Aufl. - Göttingen: Cuvillier, 2019
   Zugl.: (BTU) Cottbus-Senftenberg, Univ., Diss., 2019

   ISBN 978-3-7369-7073-1
   eISBN 978-3-7369-6073-2

Angular momentum transport and pattern formation in medium- and wide-gap turbulent Taylor-Couette flow: An experimental study

Abstract

The effective transport of momentum in turbulent flows can be further enhanced by the formation of organized flow patterns or structures. When such flows pass solid walls, a considerable drag with associated energy losses is generated. A suitable model to investigate the influence of flow patterns in turbulent flows on the drag on solid walls is the so-called Taylor-Couette (TC) system, where the fluid is confined by two coaxial and independently rotating cylinders. The TC flow features multiple different turbulent flow states, where the interaction of specific flow patterns, e.g. large-scale Taylor rolls and small-scale plumes, with the turbulence determines the angular momentum transport in the gap. This transport causes a torque on the cylinder walls due to the drag induced by the fluid. In addition, by an appropriate choice of the control parameters of the TC system, namely the shear Reynolds number Re_S, the rotation number or the ratio of angular velocities μ, and the radius ratio η, the influence of shear, rotation and wall curvature on the torque can be studied separately, which allows fundamental insights into the transport processes.

Due to the above-mentioned properties, turbulent TC flow is experimentally investigated within this thesis for medium ($\eta = 0.714$) and wide gaps ($\eta = 0.357$, $\eta = 0.5$). Direct torque measurements are performed to determine the global transport of angular momentum, while qualitative flow visualizations and quantitative velocity measurements (particle image velocimetry) in horizontal planes at different cylinder heights are used to uncover the corresponding flow patterns.

For the largely unexplored radius ratio regime of $\eta = 0.357$, the directly measured torque features a transition as a function of shear, which is connected to the capacity of the outer cylinder to emit angular momentum plumes. When the cylinders rotate slightly in counter-direction, a maximum in torque occurs at $\mu_{max}(\eta = 0.357) = -0.123$, which is induced by the formation and strengthening of large-scale Taylor vortices.

In the case of a wide-gap TC system with $\eta = 0.5$, the characteristics of the mean velocity field are analyzed in the flow regime of the torque maximum. The enhanced momentum transport at $\mu_{max}(\eta = 0.5) = -0.2$ is accompanied by a flat profile of the angular velocity in the bulk, which exemplifies the effective mixing in the gap. In addition, the contribution of the large-scale Taylor rolls to the overall momentum transport clearly exceeds the contribution of the turbulent fluctuations.

For a radius ratio of $\eta = 0.714$, the local angular momentum transport is analyzed with ($\mu_{max}(\eta = 0.714) = -0.36$) and without ($\mu = 0$) pronounced Taylor vortices. It is shown that the momentum transport due to these vortices mainly takes place at the axial height of the vortex in- and outflow regions, where the radial and azimuthal velocity components are highly correlated. This correlation results from the local ejection of small-scale plumes from the cylinder walls, which drive the large-scale circulation. Moreover, the turbulent Taylor vortices feature azimuthally traveling waves similar to the wavy Taylor vortex flow. Accordingly, the angular momentum transport in medium- and wide-gap turbulent TC flow at μ_{max} is dominated by the interaction of large-scale Taylor rolls, small-scale plumes and turbulence.

Drehimpulstransport und Strukturbildung in der turbulenten Taylor-Couette Strömung in mittleren und weiten Spalten: Eine experimentelle Studie

Zusammenfassung

Der effektive Impulstransport in turbulenten Strömungen kann durch die Ausbildung geordnete Strömungsmuster bzw. Strukturen weiter verstärkt werden. Wenn solche Strömungen feste Wände umspülen, wird ein erheblicher Reibungswiderstand und damit einhergehende Energieverluste erzeugt. Ein geeignetes Modell zur Untersuchung des Einflusses von Strömungsmustern in turbulenten Strömungen auf den Reibungswiderstand an festen Wänden ist das sogenannte Taylor-Couette (TC) System, bei dem die Flüssigkeit von zwei koaxialen und unabhängig voneinander rotierenden Zylindern begrenzt wird. Die TC-Strömung weist eine große Vielfalt turbulenter Strömungszustände auf, wobei die Wechselwirkung spezifischer Strömungsmuster, wie zum Beispiel großskaliger Taylor-Wirbel und kleinskaliger Plumes, mit der Turbulenz den Drehimpulstransport im Spalt bestimmt. Dieser Transport verursacht ein Drehmoment an den Zylinderwänden aufgrund des von der Flüssigkeit induzierten Strömungswiderstands. Darüber hinaus können der Einfluss von Scherung (Re_S), Rotation (μ) und Wandkrümmung (η) auf das Drehmoment unabhängig voneinander untersucht werden, was grundlegende Erkenntnisse über die Transportprozesse ermöglicht.

Aufgrund der oben genannten Eigenschaften wird in dieser Arbeit die turbulente TC-Strömung experimentell für mittlere ($\eta = 0.714$) und weite Spalte ($\eta = 0.357, \eta = 0.5$) untersucht. Direkte Drehmomentmessungen werden durchgeführt, um den globalen Transport des Drehimpulses zu bestimmen, während qualitative Strömungsvisualisierungen und quantitative Geschwindigkeitsmessungen (particle image velocimetry) in horizontalen Ebenen eingesetzt werden, um die entsprechenden Strömungsmuster aufzudecken.

Für das weitgehend unerforschte Radienverhältnis von $\eta = 0.357$ weist das direkt gemessene Drehmoment eine Transition abhängig von der Scherung auf, die mit dem Vermögen des äußeren Zylinders zusammenhängt, drehimpuls-transportierende Plumes zu emittieren. Wenn sich die Zylinder leicht gegenläufig drehen, tritt ein Drehmomenten-Maximum bei $\mu_{max}(\eta = 0.357) = -0.123$ auf, das durch die Ausbildung und Anfachung großskaliger Taylorwirbel erzeugt wird.

Im weiten Spalt bei $\eta = 0.5$ werden die Eigenschaften des mittleren Geschwindigkeitsfeldes im Bereich des Drehmomenten-Maximums analysiert. Der erhöhte Impulstransport bei μ_{max} geht mit einem flachen Profil der Winkelgeschwindigkeit einher, welches die effektive Durchmischung im Spalt verdeutlicht. Zudem übersteigt der Beitrag der Taylor-Wirbel am globalen Impulstransport deutlich denjenigen der Turbulenz.

Für $\eta = 0.714$ wird der lokale Impulstransport mit (μ_{max}) und ohne ($\mu = 0$) ausgeprägte Taylor-Wirbel analysiert. Es wird gezeigt, dass der Impulstransport aufgrund der Taylor-Wirbel hauptsächlich auf Höhe der Wirbelein- und -ausströmgebiete stattfindet, in denen die radiale und azimutale Geschwindigkeitskomponente deutlich korreliert sind. Diese Korrelation resultiert aus der lokalen Emission kleinskaliger Plumes von den Zylinderwänden, welche die großskalige Zirkulation antreiben. Darüber hinaus weisen die turbulenten Taylor-Wirbel Wellen auf, die sich in Umfangsrichtung ausbreiten, wie sie auch bei der welligen Taylor-Wirbelströmung auftreten. Dementsprechend wird der Dreh-impulstransport in der TC-Strömung in mittleren und weiten Spalten am Drehmomenten-Maximum durch das Zusammenwirken von großskaligen Taylor-Wirbeln, kleinskaligen Plumes und Turbulenz bestimmt.

Contents

List of Figures

List of Tables

List of Abbreviations

BL	boundary layer
BTTC	boiling Twente Taylor-Couette
CCD	charged-coupled device
CPOD	complex proper orthogonal decomposition
DNS	direct numerical simulation
EPW	end plate window
EuHIT	European High-Performance Infrastructures in Turbulence
FoV	field of view
IA	interrogation area
IC	inner cylinder
LDV	laser Doppler velocimetry
OC	outer cylinder
PDF	probability density function
PIV	particle image velocimetry
POD	proper orthogonal decomposition
RB	Rayleigh-Bénard
SVD	singular value decomposition
TC	Taylor-Couette
TP	top plate
T^{3}C	Twente turbulent Taylor-Couette
TvTCC	top-view Taylor-Couette Cottbus
2D	two-dimensional
3D	three-dimensional

List of Symbols

General conventions

X	any quantity
$X^*, X^+, \tilde{X}$	normalized quantity
X_c	complex quantity
X_{crit}	critical quantity
$\langle X \rangle_x$	averaged quantity over x

Latin letters

a	parameter of vortex extension	—
a_i	i^{th} time dependent coefficient	$\mathrm{m\,s^{-1}}$
a_{pipe}	pipe radius	m
A_E	area below co-spectra-wavenumber-curve	$\mathrm{m\,s^{-2}}$
$A_{TC,RB,pipe}$	surfaces for averages of J_{ω,T,u_z} in TC, RB and pipe	—
A_{lam}	first constant of laminar Couette solution	$\mathrm{rad\,s^{-1}}$
B_{lam}	second constant of laminar Couette solution	$\mathrm{m^2\,rad\,s^{-1}}$
$\mathbf{c} = (c_1, c_2)$	center coordinates of concentric circle fit	px
$c_{w,i}$	azimuthal wavespeed of ith CPOD mode	$\mathrm{rad\,s^{-1}}$
C	scale facor (PIV calibration)	$\mathrm{m\,px^{-1}}$
$\mathbf{CPOD}_i$	i^{th} CPOD mode	—
d	gap width	m
$\mathbf{d}_{circ,conc}$	distance vector (PIV calibration)	px
$d_{n,lam}$	effective gap width for intstability	m
d_p	diameter of a spherical tracer particle	m
$\mathbf{D}$	velocity data matrix for CPOD	$\mathrm{m\,s^{-1}}$
D_{RB}	diameter of RB cell	m
$\mathbf{e}_{r,\varphi,z}$	cylindrical unit vectors	—
$\mathbf{e}_{x,y,z}$	Cartesian unit vectors	—
$\mathbf{E}$	velocity strain tensor	$\mathrm{s^{-1}}$
E_{kin}	kinetic energy per mass	$\mathrm{m^2\,s^{-2}}$
E_{LSC}	large-scale circulation contribution to E_{kin}	$\mathrm{m^2\,s^{-2}}$
E_{turb}	turbulent contribution to E_{kin}	$\mathrm{m^2\,s^{-2}}$
E_{tot}	sum of E_{turb} and E_{LSC}	$\mathrm{m^2\,s^{-2}}$
E_{vis}	friction energy loss per mass	$\mathrm{m^2\,s^{-2}}$
$E_{r\varphi}$	azimuthal energy co-spectrum	$\mathrm{m^2\,s^{-2}}$
E_t	temporal energy spectrum	$\mathrm{m^2\,s^{-2}}$
f	frequency	Hz
$f_{1,2}$	inner, outer cylinder rotation frequenzy	Hz
$f_{w,i}$	azimuthal wave frequency of ith CPOD mode	Hz

g	gravitational acceleration	$\mathrm{m\,s^{-2}}$
$\mathbf{g}_{circ,conc}$	solution vector (PIV calibration)	px
$\mathbf{h}_{circ}$	correction vector (PIV calibration)	px
H	height of laser light sheet	m
$\mathcal{H}$	Hilbert transform of given quantity	–
H_{RB}	height of RB cell	m
$\mathbf{I}$	unit tensor	–
$\Im$	imaginary part of given quantity	–
$\mathcal{I}_{(lam)}$	light intensity (for a laminar flow)	–
$\mathcal{I}_{turb}$	turbulent contribution to $\mathcal{I}$	–
$\mathcal{I}_{LSC}$	large-scale circulation contribution to $\mathcal{I}$	–
$\mathbf{J}^{(p)}_{circ,conc}$	Jakobian matrix (PIV calibration)	–
$J_T^{(lam)}$	(laminar) heat flux	$\mathrm{m\,s^{-1}\,K}$
$J_{u_z}^{(lam)}$	(laminar) axial velocity flux	$\mathrm{m\,s^{-2}}$
$J_\omega^{(lam)}$	(laminar) angular momentum flux	$\mathrm{m^4\,s^{-2}}$
k_φ	azimuthal wavenumber vector	$\mathrm{m^{-1}}$
$k_{\varphi,w,i}$	azimuthal wavenumber of ith CPOD mode	–
K	kurtosis of a given quantity	–
K_0	modified Bessel function of the second kind	–
l	typical length	m
ℓ	cylinder length	m
$\mathcal{L}$	characteristic length	m
L	angular momentum per mass	$\mathrm{m^2\,s^{-1}}$
$L_{1,2}$	inner, outer cylinder angular momentum per mass	$\mathrm{m^2\,s^{-1}}$
L_{pipe}	length of pipe	m
$n_{1,2}$	inner, outer cylinder rotation rate	rpm
n_{TV}	number of Taylor vortices	–
$N_{r,\varphi,z}$	number of grid points for DNS	–
p	pressure	$\mathrm{kg\,m^{-1}\,s^{-2}}$
$p_{1,2,3}$	quadratic fit parameter for Nu_ω as function of μ	–
(r,φ,z)	cylindrical coordinates	m, rad, m
$r_{1,2}$	inner, outer cylinder radius	m
$r_{n,(lam)}$	position of neutral line (for a laminar flow)	m
$r_{n,EG}$	position of neutral line with vortex extension	m
$r_{p,EG}$	position of instability inset	m
$r_{pic,1,2}$	inner, outer cylinder image radii	px
$\Re$	real part of given quantity	–
R_{rr}	azimuthal two-point auto-correlation of u_r	–
$R_{\varphi\varphi}$	azimuthal two-point auto-correlation of u_φ	–
$R_{\mathcal{I}\mathcal{I}}$	axial two-point auto-correlation of $\langle \mathcal{I} \rangle_t$	–
s	arc length	m
s_{pic}	constant for IC detection	px
S	skewness of a given quantity	–
t	time coordinate	s
T	temperature	K
$\mathbf{T}$	stress tensor	$\mathrm{kg\,m^{-1}\,s^{-2}}$

$\mathcal{T}_{(pa)}$	(progressing average of) torque	$\mathrm{kg\,m^2\,s^{-2}}$
$\mathbf{u} = (u_x, u_y, u_z)$	inertial frame Cartesian velocity vector	$\mathrm{m\,s^{-1}}$
$\mathbf{u} = (u_r, u_\varphi, u_z)$	inertial frame cylindrical velocity vector	$\mathrm{m\,s^{-1}}$
$\mathbf{u}_{pic}$	inertial frame image Cartesian velocity vector	px
u_S	shear velocity	$\mathrm{m\,s^{-1}}$
$u_{\varphi,1,2}$	inner, outer cylinder azimuthal velocity	$\mathrm{m\,s^{-1}}$
$u_{\tau,(1,2)}$	(inner, outer cylinder) frictional velocity	$\mathrm{m\,s^{-1}}$
$\mathcal{U}$	characteristic velocity	$\mathrm{m\,s^{-1}}$
$\hat{\mathcal{U}}$	Fourier transform of velocity	$\mathrm{m\,s^{-1}}$
U_{pipe}	cross-sectional averaged velocity in pipe	$\mathrm{m\,s^{-1}}$
U_W	typical amplitude of wind	$\mathrm{m\,s^{-1}}$
$\mathbf{w} = (w_r, w_\varphi, w_z)$	rotating frame cylindrical velocity vector	$\mathrm{m\,s^{-1}}$
(x, y, z)	Cartesian coordinates	m
$\mathbf{x}$	position vector	m
$\mathbf{x}_{pic} = (x_{pic}, y_{pic})$	image Cartesian position vector	px
$\mathbf{x}_{pic,1,2}$	inner, outer cylinder image Cartesian coordinates	px

Greek letters

α	exponent of scaling law between Nu_ω and Re_S	–
α_{EG}	constant of viscous stability criterion	–
α_T	thermal expansion coefficient of the fluid	$\mathrm{K^{-1}}$
β	exponent of scaling law between Re_W and Re_S	–
γ	exponent of scaling law between $E_{r\varphi}$ and k_φ	–
Γ	aspect ratio	–
δ	infinitesimal value of a given quantity	–
$\delta_{v,(1,2)}$	(inner, outer cylinder) viscous length scale	m
$\triangle$	measurement uncertainty of a given quantity	–
Δ	increment of a given quantity	–
Δ_{EG}	cross-over function of viscous stability criterion	–
Δ_{mean}	resolvable displacement of averaged PIV	–
Δ_{SM}	resolvable displacement of single PIV	–
Δ_{10}	bin size in logarithmic scale	–
$\epsilon_{(lam)}$	(laminar) energy dissipation rate per mass	$\mathrm{m^2\,s^{-3}}$
ϵ_W	wind energy dissipation rate per mass	$\mathrm{m^2\,s^{-3}}$
η	radius ratio	–
η_v	dynamic viscosity of the fluid	$\mathrm{kg\,m^{-1}\,s^{-1}}$
κ	thermal diffusivity of the fluid	$\mathrm{m^2\,s^{-1}}$
$\lambda_{BL,1,2}$	inner, outer cylinder BL thickness	m
λ_{Laser}	wavelength of Laser light	m
λ_{TV}	axial wavelength of Taylor vortices	–
$\lambda_{w,i}$	azimuthal wavelength of ith CPOD mode	rad
μ	ratio of angular velocities	–
μ_{max}	ratio of angular velocities at torque maximum	–
μ_b	angle bisector prediction for μ_{max}	–
μ_p	Brauckmann/Eckhardt prediction for μ_{max}	–
ν	kinematic viscosity of the fluid	$\mathrm{m^2\,s^{-1}}$

ρ	fluid density	$\mathrm{kg\,m^{-3}}$
ρ_p	density of a tracer particle	$\mathrm{kg\,m^{-3}}$
ρ_P	correlation coefficient	$-$
σ	standard deviation	$-$
Σ	matrix of singular values (SVD)	$-$
τ_f	fluid time scale	s
τ_p	particle resonse time	s
τ_W	wall shear stress	$\mathrm{kg\,m^{-1}\,s^{-2}}$
τ_ν	viscous time	s
ϕ_i	i^{th} temporal phase function	$-$
Φ	matrix of left singular vectors (SVD)	$-$
Φ_i	i^{th} spatial phase function	$-$
Ψ	stream function	$\mathrm{m^2\,s^{-1}}$
$\boldsymbol{\Psi}$	matrix of right singular vectors (SVD)	$-$
$\omega_{1,2}$	inner, outer cylinder angular velocity	$\mathrm{rad\,s^{-1}}$
$\omega_{(lam)}$	(laminar) angular velocity	$\mathrm{rad\,s^{-1}}$
Ω	spin tensor	$\mathrm{s^{-1}}$
Ω_{rf}	rotating frame angular velocity	$\mathrm{rad\,s^{-1}}$

Math operators

d	total derivative
D_t	material derivative
∂	partial derivative
∇	nabla operator

Dimensionless numbers

G	dimensionless torque
Nu_T	heat Nusselt number
Nu_{u_z}	pipe Nusselt number
Nu_ω	pseudo-Nusselt number
$\mathrm{Nu}_\omega^{turb}$	turbulent contribution to Nu_ω
Nu_ω^{LSC}	large-scale circulation contribution to Nu_ω
$\mathrm{Nu}_{\omega,stress,grad}^{LSC}$	stress, gradient fraction of Nu_ω^{LSC}
Nu_ω^{v}	viscous contribution to Nu_ω
$\mathrm{Nu}_\omega^{c,(net)}$	(net) convective contribution to Nu_ω
$\mathrm{Nu}_{\omega,In}^{c,net}$	net convective contribution at vortex inflow to Nu_ω
$\mathrm{Nu}_{\omega,Out}^{c,net}$	net convective contribution at vortex outflow to Nu_ω
Pr	Prandtl number
Ra	Rayleigh number
R_C	curvature number
Re	Reynolds number
$Re_{1,2}$	inner, outer cylinder Reynolds number
$Re_{\tau,1,2}$	inner, outer cylinder frictional Reynolds number
Re_{pipe}	pipe Reynolds number
Re_S	shear Reynolds number

Ro	Rossby number
R_Ω	rotation number
St	Stokes number
Ta	Taylor number
σ_{Pr}	pseudo-Prandtl number

*Dedicated to
my beloved wife
Stefanie
and my son
Felix.*

Chapter 1

Motivation

1.1 Turbulence in fluid motion

Most of the fluid motions in nature like rivers, oceans or the atmosphere, and in technology as the flow around cars, ships or airplanes are turbulent. Turbulence is characterized by strong spatial and temporal fluctuations of the velocity along with eddies of many scales. Thus, the flow field appears irregular and chaotic, which is why the turbulent fluid motion is hard to predict [93, 102]. These properties induced the physicist Sir Horace Lamb to make the following statement:

> *"I am an old man now, and when I die and go to heaven, there are two matters on which I hope for enlightenment. One is quantum electrodynamics and the other is the turbulent motion of fluids. About the former, I am really rather optimistic."*

A flow becomes turbulent when the inertia forces sufficiently exceed the viscous forces. Their ratio yields the so-called Reynolds number $Re = \mathcal{U}\mathcal{L}/\nu$, where $\mathcal{U}$ and $\mathcal{L}$ are, respectively, a characteristic velocity and length scale of the flow, and ν is the kinematic viscosity of the fluid [101]. When Re is small, velocity fluctuations are damped by viscosity, and the fluid moves unidirectional without notable mixing between adjacent fluid layers. Above a critical Reynolds number, the flow transitions to a turbulent state, including complex vortical fluid motion. Two well-known instability mechanisms driving a transition are the shear instability [107], where the velocity difference of neighboring fluid layers reaches a critical value, and the centrifugal instability [121], where the centrifugal force sufficiently exceeds the counter-acting pressure gradient.

Turbulent flows are of special interest for engineers, as they transport and mix matter, momentum and heat much more effectively than laminar flows. This characteristic is of great importance for flows in the vicinity of solid walls, where the fluid velocity vanishes. Due to turbulent mixing, the momentum of the free stream velocity is strongly transported into the near-wall region, the so-called boundary layer (BL), and the resulting large velocity gradient at the wall causes an enhanced drag force [93]. Accordingly, the understanding of turbulence can help to reduce friction losses and therefore fuel consumption of cars, airplanes or turbomachines.

Beside inertia and viscosity, other forcing like stratification [9, 142] or magnetic fields [103] can have an impact on the flow stability. Therefore, complex turbulent flows are often investigated in simple geometries, where the influence of only a few driving parameters can be analyzed separately. Such a geometry is the so-called Taylor-Couette flow, which is the subject of this thesis.

1.2 Taylor-Couette (TC) flow

The flow in the gap between two coaxial and independently rotating cylinders is called the Taylor-Couette (TC) flow (see Figure 1.1(a)). It is a widely used setup in fluid dynamics research as a model for wall-bounded rotating shear flows and has many technical applications as rotating machinery, journal bearings or centrifuges [67, 108, 119], and also has astrophysical relevance concerning accretion discs [6, 31]. In the classical approach, two Reynolds numbers are defined for the inner (IC, index 1) and outer cylinder (OC, index 2). Using the cylinder radii $r_{1,2}$, the cylinder angular velocities $\omega_{1,2}$ and the gap width $d = r_2 - r_1$, these Reynolds numbers are defined by $Re_{1,2} = r_{1,2}\omega_{1,2}d/\nu$. Dubrulle *et al.* [31] proposed two alternative dimensionless parameters from a dynamical point of view, leading to the shear Reynolds number $Re_S = 2|Re_1 - \eta Re_2|/(1+\eta)$ and the rotation number $R_\Omega = (1-\eta)(Re_1 + Re_2)/(\eta Re_2 - Re_1)$, which is a function of the ratio of angular velocities $\mu = \omega_2/\omega_1$. This choice of parameters enables to investigate the influence of shear and rotation separately. In addition, the fluid moves along curvilinear cylinder surfaces, which introduces curvature effects. As a measure for the curvature, the ratio of the cylinder radii is used: $\eta = r_1/r_2$. When η approaches to 1, no curvature is experienced by the flow [17]. Furthermore, the gap is enclosed by two endplates at the length ℓ, which translates into an aspect ratio of $\Gamma = \ell/d$.

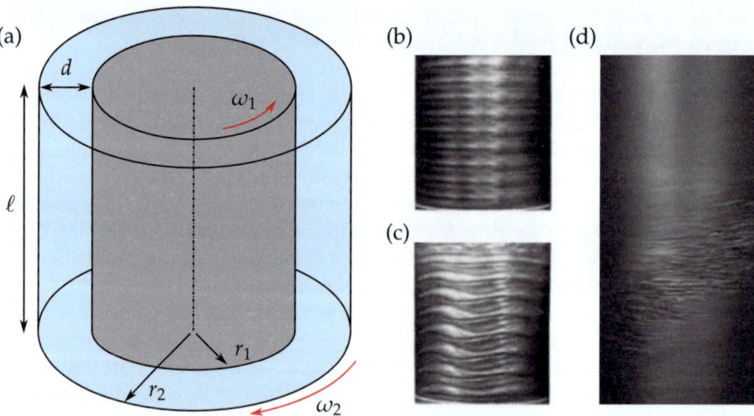

FIGURE 1.1: (a) Sketch of a Taylor-Couette system, consisting of an inner (1) and outer cylinder (2). r represents the radius, d the gap width, ℓ the length and ω the angular velocity of the cylinders. (b) Snapshots of the supercritical Taylor vortex flow and (c) wavy Taylor vortex flow, taken from Fardin *et al.* [39]. (d) Snapshot of a subcritical turbulent spiral arm, taken from Avila and Hof [4]

The popularity of the TC setup has many reasons [51]. It is a closed system, where the global energy input due to the cylinder rotation and the energy dissipation inside the gap are exactly balanced [32]. Moreover, the geometry is simple and periodic into the azimuthal flow direction. If stable flow features like coherent structures exist in the gap, they are not carried away downstream and can be observed for infinite time. Besides, the cylinder speeds are precisely controllable, and the system features two boundary layers of different size when $\eta < 1$ [33]. The latter fact makes the TC flow appropriate to investigate the interaction of boundary layers and the bulk flow.

A unique feature of TC flows is the existence of two different transition scenarios to turbulence. The laminar baseflow, which is purely azimuthal, is called Couette flow. When only the inner cylinder rotates and reaches a critical value, the Couette flow becomes centrifugally unstable, and axially regular spaced, toroidal vortices are formed, the so-called Taylor vortices (Taylor [121], see Figure 1.1(b)). The axial number of these vortices is non-unique and depends on the initial flow conditions and the acceleration rate of the cylinder [116]. At even higher cylinder speeds, two further instabilities with modulations of the flow in space and time appear before the flow becomes fully turbulent (Fenstermacher *et al.* [40], see Figure 1.1(c)). This route into chaos is called a supercritical transition, that is similar to the one found in the Rayleigh-Bénard (RB) flow [21]. There, a fluid layer is heated from below and cooled from above.

When the outer cylinder rotation is dominant, the flow directly transitions to turbulence at a sufficiently high Reynolds number due to a formation of turbulent spots or spirals [23, 48]. These localized turbulent regions exist next to the laminar baseflow (see Figure 1.1(d)) and grow with increasing Reynolds number until the whole gap becomes turbulent. Such a scenario with spatio-temporal intermittency is called subcritical transition and is also observable in pipe flow, the pressure-driven flow through a circular tube [34, 54]. Moreover, multiple different flow states are adjustable in the transitional Reynolds number regime for independently rotating cylinders [3, 23]. Similarly, Ostilla-Mónico *et al.* [86] reported manifold flow states also in the turbulent regime. Prescribing Re_S and μ, turbulent Taylor vortices either capture the whole gap, are restricted to an inner gap region with an intermittent outer region, or vanish leading to featureless turbulence. Furthermore, a strong connection between large-scale turbulent Taylor rolls and small-scale structures, which are emitted from the cylinder walls, has been identified by Ostilla-Mónico *et al.* [83, 86]. These small-scale structures, namely hairpin vortices or plumes in the context of RB flows, transport angular momentum and contribute to the large-scale circulation in the gap. The variety of transitional and turbulent flow states, occurring in TC geometry and including large- and small-scale flow patterns, make the TC flow a paradigmatic system in fluid dynamics research to investigate flow instabilities [23, 26, 121], pattern formation [3, 12] and turbulence [51, 60, 66, 130].

In between the cylinder walls, the radial transport of angular momentum J_ω is conserved [33]. This quantity is proportional to the torque $\mathcal{T}$ and therefore to the drag force the fluid is acting on the cylinder walls, and to the energy dissipation rate ϵ. Accordingly, J_ω is the most important response parameter in TC flows, as it is a measure for friction losses inside the gap. In the 18th century, Couette [24] and Mallock [70] measured the torque in a TC apparatus adjusting a laminar flow to quantify the viscosity of fluids. Henceforth, the momentum transport has been widely investigated experimentally [66, 68, 78, 132] as well as numerically [14, 15, 28, 29, 83, 86, 87] as a function of shear (Re_S), rotation (μ) and curvature (η) in the fully turbulent regime. J_ω scales nearly exponentially with Re_S for pure inner cylinder rotation and depicts a maximum for slight counter-rotation. The maximum location μ_{max} depends on η and is caused by strengthened turbulent Taylor vortices [15]. In addition, these large-scale vortices consist and are driven by the small-scale plumes [83, 86]. This interaction between structures of different scales, that highly influence the momentum transport, makes a detailed analysis of J_ω

very hard to realize. The aim of the current thesis is to provide novel insights into the momentum transport and the underlying flow structure.

1.3   Aim and outline of the thesis

The investigation of the radial angular momentum transport in the presence of turbulent Taylor vortices and small-scale plumes is a challenge for numerical simulations as well as for experiments. Due to the existence of large-scale coherent structures in the fully turbulent regime, the flow statistics clearly depend on all three coordinate directions [60, 128]. Numerical simulations provide the torque at the cylinder wall and the three-dimensional flow field, which enables a combined analysis of momentum transport and the local flow organization. However, the computational costs strongly increase with the Reynolds number as $\approx Re^4$ [93], which limits the maximum achievable forcing to relatively low Re, with only a few exceptional studies reaching $Re = \mathcal{O}(10^4 - 10^5)$ [15, 17, 86, 88]. Aside, experiments in TC flows can achieve Reynolds numbers up to $Re = \mathcal{O}(10^6)$ and precisely determine the torque [66, 78, 132], but quantitative measurements of the full velocity field are difficult to realize. Mostly, particle-based optical measurement techniques like laser Doppler velocimetry (LDV) or particle image velocimetry (PIV) are used [55, 139, 140]. While LDV only provides point-wise velocity information, PIV can capture the velocity field within sheets [133] or in a volume [123, 124]. However, optical distortions due to the curved rotating cylinder surfaces and changes in the refractive index complicate the measurements. Further, volumetric PIV is strongly limited in the achievable spatial resolution, as shown by Tokgoz *et al.* [124]. This fact is especially important due to the pronounced interaction of the large-scale rolls with small-scale plumes. To overcome these experimental limitations within this thesis, the velocity field is measured through a plane top plate in horizontal sheets at different cylinder heights, which was successfully used already by van der Veen *et al.* [128]. This configuration prevents refraction on the curved OC and offers a sufficient spatial resolution as well as quasi-three-dimensional velocity data. However, the proposed PIV technique is restricted to the analysis of statistical quantities without instantaneous information on coherent structures, concerning the axial coordinate direction. In combination with direct torque measurements and flow visualizations, the momentum transport and the underlying flow structures can be investigated. Further, within the literature, most of TC studies concentrate on wide, medium and narrow gaps ($\eta \geq 0.5$), while investigations for even wider gaps ($\eta < 0.5$) are very rare. As the momentum transport and especially its maximum location strongly depend on η, the parameter space of torque measurements is extended within this thesis to the unexplored regime of $\eta = 0.357$. Further, for $\eta = 0.5$ and $\eta = 0.714$, where the angular momentum transport has already been investigated, the velocity field is analyzed in detail to uncover the contribution of large-scale Taylor rolls to J_ω and their interaction with small-scale plumes at high Reynolds numbers. In summary, the following scientific questions are addressed within this thesis: (*i*) How does the angular momentum scales with shear and rotation for a wide gap of $\eta = 0.357$, (*ii*) to what extent do the large-scale Taylor rolls contribute to the overall momentum transport in the fully turbulent regime, (*iii*) how do the large-scale Taylor rolls influence the characteristics of the mean velocity field and (*iv*) what role does the interaction of small-scale plumes and large-scale Taylor rolls play for the momentum transport.

The thesis is organized as follows. The fundamental equations of motion are summarized in Chapter 2. Therefore, the Navier-Stokes equation with TC specific boundary conditions is introduced to derive the dimensionless key parameters. In addition, the laminar Couette baseflow, the centrifugal instability leading to the Taylor vortex flow and the exact relation of the angular momentum transport are deduced. In the end, the analogy between TC flow, pipe flow and RB flow, as well as the main parameters used throughout the thesis are highlighted.

Chapter 3 provides an overview of the current state of research in turbulent TC flow with the main focus on the angular momentum transport in medium and wide gaps. In the beginning, the effective scaling of the torque with the shear Reynolds number is discussed, where no pure power-law scaling is found. Subsequently, the dependence of $\mathcal{T}$ on the rotation ratio is described, and two predictions for the torque maximum location are reviewed. Further, velocity field properties, characteristics of turbulent Taylor vortices and their contribution to the global momentum transport are presented. Finally, the interaction of large-scale turbulent Taylor vortices and small-scale plumes, as well as the influence of curvature on TC flows are discussed.

The experimental facilities used within this thesis are described in Chapter 4. The setup of the top-view Taylor-Couette Cottbus experiment (TvTCC), its control and the used working fluids are specified in detail. Moreover, the boiling Twente Taylor-Couette facility (BTTC) from the University of Twente is roughly outlined, where a measurement campaign took place within the framework of an European High-Performance Infrastructures in Turbulence (EuHIT) project.

Chapter 5 focuses on the measurement techniques used for the investigations. Qualitative information on the flow organization are captured with a particle-based global flow visualization method. The radial transport of angular momentum is determined by direct torque measurements over the whole length of the inner cylinder. To validate the torque signal, measurements for the well-known radius ratio of $\eta = 0.5$ are compared with existing numerical and experimental data. To analyze the velocity field, planar PIV is performed in horizontal planes at different cylinder heights. Special attention is paid to the calibration of this technique, and possible measurement errors are discussed.

The Chapters 6,7 and 8 contain the experimental results of this thesis. In Chapter 6, direct torque measurements and flow visualizations are performed for a radius ratio of $\eta = 0.357$, which are unique in such a wide-gap configuration. The effective scaling of the torque with Re_S, the location of the torque maximum and the underlying flow organization are analyzed. To strengthen the findings, the experimental results are compared to direct numerical simulations (DNSs), provided by Rodolfo Ostilla-Mónico.

In Chapter 7 height dependent PIV measurements are shown for a radius ratio of $\eta = 0.5$. Mean flow statistics concerning the radial profiles of angular velocity, the energy distribution and the strength of the secondary flow are analyzed in the region of the torque maximum and for different shear rates. Further, the contribution of the large-scale turbulent Taylor vortices to the overall angular momentum transport is worked out. Note that in this and in the next chapter no direct torque measurements were performed and the transport-related findings are purely based on velocity data.

In Chapter 8 small-scale statistics of TC flows for $\eta = 0.714$ are investigated, again using height dependent PIV. The flow field for pure inner cylinder rotation ($\mu = 0$) with featureless turbulence and for μ_{max} with pronounced turbulent Taylor vortices are compared. The contribution of the vortex in- and outflow to the momentum transport and the interaction of the large-scale vortices and small-scale structures are analyzed based on mean field statistics, probability density functions (PDFs), velocity spectra and two-point correlations. Further, azimuthally traveling waves superimposed on the turbulent Taylor vortices are analyzed using a complex proper orthogonal decomposition (CPOD). In the end, the findings of the current study are summarized and an outlook is given in Chapter 9.

The results of this thesis are mainly based on three papers: (*i*) A. Froitzheim, S. Merbold, R. Ostilla-Mónico and C. Egbers (2019), *Angular momentum transport and flow organization in Taylor-Couette flow at radius ratio of* $\eta = 0.357$, Phys. Rev. Fluids 4 (2019), 084605 [41]; (*ii*) A. Froitzheim, S. Merbold and C. Egbers, *Velocity profiles, flow structures and scalings in a wide-gap turbulent Taylor-Couette flow*, J. Fluid Mech. 831 (2017), 330-357 [45]; and (*iii*) A. Froitzheim, R. Ezeta, S.G. Huisman, S. Merbold, C. Sun, D. Lohse and C. Egbers, *Statistics, plumes and azimuthally traveling waves in ultimate Taylor-Couette turbulent vortices*, J. Fluid Mech. 876 (2019), 733-765 [43].

Chapter 2

Fundamentals

2.1 Cylindrical coordinate system

The Taylor-Couette geometry consists of two coaxial cylinders. Thus, the motion of the flow is naturally described in the cylindrical instead of Cartesian coordinates. The corresponding transformation is shown in Figure 2.1 [113]. Bold symbols represent vectors.

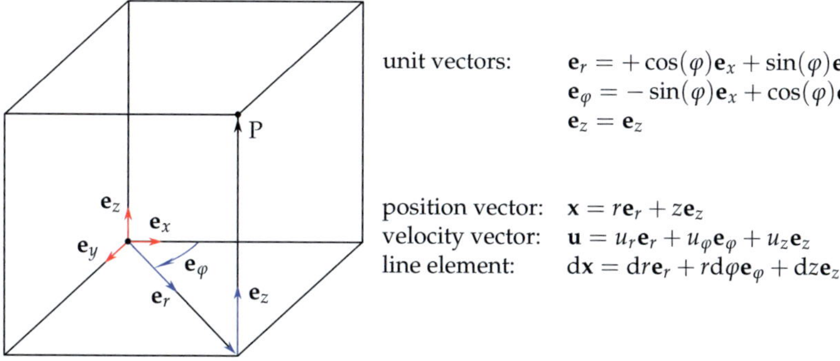

$$\mathbf{e}_r = +\cos(\varphi)\mathbf{e}_x + \sin(\varphi)\mathbf{e}_y$$
$$\mathbf{e}_\varphi = -\sin(\varphi)\mathbf{e}_x + \cos(\varphi)\mathbf{e}_y$$
$$\mathbf{e}_z = \mathbf{e}_z$$

$$\mathbf{x} = r\mathbf{e}_r + z\mathbf{e}_z$$
$$\mathbf{u} = u_r\mathbf{e}_r + u_\varphi\mathbf{e}_\varphi + u_z\mathbf{e}_z$$
$$d\mathbf{x} = dr\mathbf{e}_r + rd\varphi\mathbf{e}_\varphi + dz\mathbf{e}_z$$

FIGURE 2.1: Sketch of transformation from cylindrical to Cartesian coordinates, adopted from Spurk and Aksel [113]. Corresponding equations are depicted on the right-hand side.

Of special interest is the conversion of the Cartesian velocity components into the cylindrical ones, which becomes based on the above-mentioned transformation:

$$u_r = +u_x \cos(\varphi) + u_y \sin(\varphi), \tag{2.1}$$
$$u_\varphi = -u_x \sin(\varphi) + u_y \cos(\varphi). \tag{2.2}$$

In the following, these cylindrical coordinates and velocities are used for the equations of motion.

2.2 Equations of motion

To derive the equations of motion of a fluid, the continuum hypothesis has to be introduced. In general, it is not possible to follow the motion of individual molecules due to

Heisenberg's Uncertainty Principle. Instead, an infinitely small cluster of molecules is assumed as the smallest part of the material, whose occupied volume is small compared to the macroscopic length of interest. This cluster is called fluid particle [113]. In addition, its volume has to be large enough, that its properties can be averaged over the volume, independent on the number of molecules inside the volume. If this assumption holds, the properties of the fluid become a continuous function of space and time and the fluid itself can be treated as a continuum. The fluid particle is further considered as a material point. The following explanations are based on Spurk and Aksel [113].

2.2.1 Kinematics of a fluid particle

The acceleration, a fluid particle encounters, when passing through a point $\mathbf{x}$ at time t, consists of a local and a convective change on its way from $\mathbf{x}$ to $\mathbf{x} + d\mathbf{x}$. Therefore, the acceleration is

$$D_t\mathbf{u} = \partial_t\mathbf{u} + (\mathbf{u} \cdot \nabla)\,\mathbf{u}, \qquad \nabla = \begin{bmatrix} \partial_r \\ \dfrac{1}{r}\partial_\varphi \\ \partial_z \end{bmatrix}. \tag{2.3}$$

The velocity vector $\mathbf{u}(\mathbf{x}, t)$ is a function of space $\mathbf{x}$ and time t and the operator D_t is called material derivative with $D_t = \partial t + \mathbf{u} \cdot \nabla$. It is important to note that the convective term in equation (2.3) is non-linear, as the products of the velocity with its first derivative appear. If the velocity at the position $\mathbf{x}$ is known, the velocity at a nearby position $\mathbf{x} + d\mathbf{x}$ can be calculated using the Taylor expansion:

$$\mathbf{u}\,(\mathbf{x} + d\mathbf{x}, t) = \mathbf{u}\,(\mathbf{x}, t) + d\mathbf{x} \cdot \nabla\mathbf{u}. \tag{2.4}$$

$\nabla\mathbf{u}$ is the velocity gradient tensor, which is a second order tensor:

$$\nabla\mathbf{u} = \begin{bmatrix} \partial_r u_r & \dfrac{1}{r}\left(\partial_\varphi u_r - u_\varphi\right) & \partial_z u_r \\ \partial_r u_\varphi & \dfrac{1}{r}\left(\partial_\varphi u_\varphi + u_r\right) & \partial_z u_\varphi \\ \partial_r u_z & \dfrac{1}{r}\partial_\varphi u_z & \partial_z u_z \end{bmatrix}. \tag{2.5}$$

The velocity gradient tensor can be decomposed into a symmetric and an anti-symmetric tensor in such a way, that equation (2.4) becomes

$$\mathbf{u}\,(\mathbf{x} + d\mathbf{x}, t) = \mathbf{u}\,(\mathbf{x}, t) + d\mathbf{x} \cdot \mathbf{E} + d\mathbf{x} \cdot \Omega, \tag{2.6}$$

$$\mathbf{E} = \frac{1}{2}\left(\nabla\mathbf{u} + (\nabla\mathbf{u})^T\right), \tag{2.7}$$

$$\Omega = \frac{1}{2}\left(\nabla\mathbf{u} - (\nabla\mathbf{u})^T\right). \tag{2.8}$$

The first term on the right-hand side of equation (2.6) denotes a translation, the second term a deformation and the third one a solid body rotation. Therefore, $\mathbf{E}$ is called rate of deformation tensor or velocity strain tensor and $\mathbf{\Omega}$ spin tensor. As friction stresses in a fluid do only arise in the presence of a deformation, they just depend on $\mathbf{E}$ and not on $\mathbf{\Omega}$. Therefore, the diagonal elements of the velocity strain tensor can be interpreted as stretching velocities parallel to the axis of a fluid element and the non-diagonal elements as shear velocities [113].

2.2.2 The continuity equation

The continuity equation is based on the conservation of mass [113]. In a fluid volume, the sum of all positive and negative mass fluxes has to be equal to the local change of the density $\rho(\mathbf{x}, t)$:

$$D_t \rho + \rho \nabla \cdot \mathbf{u} = 0. \tag{2.9}$$

If the density of a fluid particle is constant along its path of motion, the flow is preserving its volume and is called incompressible. Most fluids satisfy this assumption, which means that $D_t \rho = 0$. Therefore, equation (2.9) simplifies to

$$\nabla \cdot \mathbf{u} = 0. \tag{2.10}$$

2.2.3 The Navier-Stokes equation in an inertial frame

According to the first axiom of classical mechanics, the change of momentum of a body is equal to the force acting on this body [113]. Here, the body is a fluid particle with the specific momentum $\rho \mathbf{u}$. The forces acting on the body can be of two kinds, namely body forces and contact forces. If only the gravitational force $\mathbf{g} = g \mathbf{e}_z$ is assumed as a body force, which is valid in an inertial frame of reference without other external forces, the balance of momentum of a fluid particle is given by

$$\rho D_t \mathbf{u} = \rho \mathbf{g} + \nabla \cdot \mathbf{T}. \tag{2.11}$$

Equation (2.11) is known as Cauchy's first law of motion and is valid for each kind of continuum. $\mathbf{T}$ represents the stress tensor (second order) containing normal stresses as diagonal elements and shear stresses as non-diagonal elements. To solve equation (2.11), a material law of the fluid is needed. If a linear relationship between the components of the stress tensor $\mathbf{T}$ and the velocity strain tensor $\mathbf{E}$ as well as an incompressible fluid are assumed, the stress tensor becomes

$$\mathbf{T} = -p\mathbf{I} + 2\eta_v \mathbf{E}. \tag{2.12}$$

$p(\mathbf{x}, t)$ represents the dynamic pressure, which is in case of an incompressible fluid independent on the thermodynamic state of the fluid. $\mathbf{I}$ is the unit tensor and η_v the dynamic viscosity of the fluid, which is related to the kinematic viscosity v by $v = \eta_v / \rho$. The linear relationship shown in equation (2.12) is valid for so-called Newtonian fluids, which includes most of the gases and liquids of low molecular weight as air or water. Inserting the material law of Newtonian incompressible fluids into Cauchy's first law of motion yields the Navier-Stokes equation. If further the gravitational force is neglected, the Navier-Stokes equation writes as

$$\partial_t \mathbf{u} + (\mathbf{u} \cdot \nabla)\,\mathbf{u} = -\frac{1}{\rho}\nabla p + \nu\nabla^2\mathbf{u}. \tag{2.13}$$

Concerning the Taylor-Couette geometry, the boundary conditions of the velocity are defined as $\mathbf{u}\,(r_1) = [0, r_1\omega_1, 0]^T$ and $\mathbf{u}\,(r_2) = [0, r_2\omega_2, 0]^T$. The exponent T represents the transpose of the vector. Using the gap width $d = r_2 - r_1$ as characteristic length scale and $\tau_\nu = d^2/\nu$, the so-called viscous time, as characteristic time scale (see Dubrulle *et al.* [31]), equation (2.13) becomes for the dimensionless velocity $\mathbf{u}^*$

$$\partial_t \mathbf{u}^* + (\mathbf{u}^* \cdot \nabla)\,\mathbf{u}^* = -\nabla p^* + \nabla^2\mathbf{u}^*, \tag{2.14}$$

with the boundary conditions $\mathbf{u}^*\,(r_1) = [0, Re_1, 0]^T$ and $\mathbf{u}^*\,(r_2) = [0, Re_2, 0]^T$. Here, the classical inner and outer cylinder Reynolds number come into play:

$$Re_1 = \frac{r_1\omega_1 d}{\nu}, \qquad Re_2 = \frac{r_2\omega_2 d}{\nu}. \tag{2.15}$$

2.2.4 The Navier-Stokes equation in a rotating frame

When the axes of the cylindrical coordinate system are accelerated and not fixed in space, equation (2.13) is not valid anymore. Within this section, a relative system is considered, whose origin is at rest and rotates with a constant angular velocity $\Omega_{rf} = \Omega_{rf}\mathbf{e}_z$ according to Figure 2.2. With the absolute velocity $\mathbf{u}$ in the inertial frame (index I) and the relative velocity $\mathbf{w}$ in the rotating frame (index R), their dependency is given by

$$\mathbf{u} = \mathbf{w} + \Omega_{rf} \times \mathbf{x}. \tag{2.16}$$

Thus, the acceleration in the rotating frame is related to the acceleration in the inertial frame as

$$[D_t\mathbf{u}]_I = [D_t\mathbf{w}]_R + 2\Omega_{rf} \times \mathbf{w} + \Omega_{rf} \times (\Omega_{rf} \times \mathbf{x}). \tag{2.17}$$

The two additional terms, which appear in equation (2.17), are the Coriolis force and the centrifugal force. As these forces are only felt by a body in the rotating frame, they are called apparent forces and can be classified as body forces. When the centrifugal force is formulated as $\Omega_{rf} \times (\Omega_{rf} \times \mathbf{x}) = -0.5\nabla\,(\Omega_{rf} \times \mathbf{x})^2$, it becomes obvious, that this apparent force only shifts the pressure in the system and can be included into p. Further, an incompressible, Newtonian fluid is assumed and the gravitational force is neglected, which leads to the Navier-Stokes equation in a rotating frame [31]

$$\partial_t \mathbf{w} + (\mathbf{w} \cdot \nabla)\,\mathbf{w} = -\frac{1}{\rho}\nabla p + \nu\nabla^2\mathbf{w} - 2\Omega_{rf} \times \mathbf{w}, \tag{2.18}$$

with the Taylor-Couette specific boundary conditions $\mathbf{w}(r_1) = [0, r_1(\omega_1 - \Omega_{rf}), 0]^T$ and $\mathbf{w}(r_2) = [0, r_2(\omega_2 - \Omega_{rf}), 0]^T$. Here, the question arises how to properly choose the angular velocity Ω_{rf} in TC geometry. To rebuild the symmetry between the inner and outer cylinder wall velocities, Ω_{rf} is set to fulfill the equation $w_\varphi(r_1) = -w_\varphi(r_2)$ in the rotating frame, according to Dubrulle *et al.* [31]. Thus, it follows

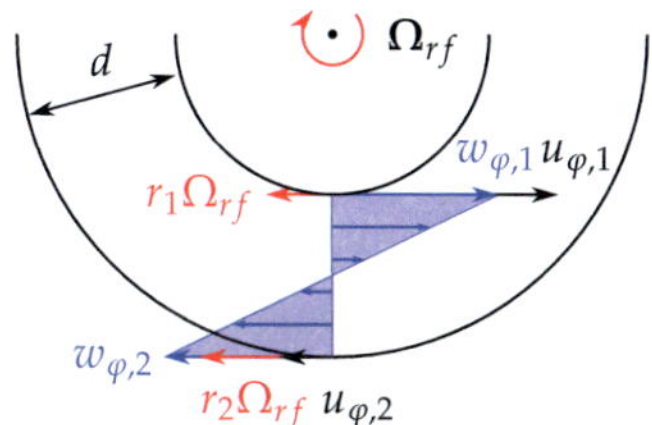

FIGURE 2.2: Sketch of the relation between absolute and relative velocities in a reference frame rotating with Ω_{rf} (modified from [17]).

$$\Omega_{rf} = \frac{r_1\omega_1 + r_2\omega_2}{r_1 + r_2}. \tag{2.19}$$

To normalize equation (2.18), d is used again as characteristic length scale and the velocity difference of the cylinder walls u_S, which is called the shear velocity, as characteristic velocity scale

$$u_S = \left| w_\varphi\left(r_1\right) - w_\varphi\left(r_2\right) \right| = \frac{2}{\eta + 1}\left|r_1\omega_1 - \eta r_2\omega_2\right|, \tag{2.20}$$

with the radius ratio $\eta = r_1/r_2$. The normalized Navier-Stokes equation in a rotating frame follows as

$$\frac{\partial \mathbf{w}^*}{\partial t^*} + \left(\mathbf{w}^* \cdot \nabla\right)\mathbf{w}^* = -\nabla p^* - R_\Omega \mathbf{e}_z \times \mathbf{w}^* + \frac{1}{Re_S}\nabla^2 \mathbf{w}^*, \tag{2.21}$$

with the boundary conditions

$$w_\varphi^*(r_1) = u_\varphi^*(r_1) - \frac{\eta}{2(1-\eta)}R_\Omega, \quad w_\varphi^*(r_2) = u_\varphi^*(r_2) - \frac{1}{2(1-\eta)}R_\Omega. \tag{2.22}$$

According to equation (2.21), the flow field is described by two dimensionless numbers, namely the shear Reynolds number Re_S, which is a measure for the shear rate, and the rotation number R_Ω, measuring the effect of rotation on the flow. The second quantity can be described in terms of the ratio of angular velocities $\mu = \omega_2/\omega_1$, which is called the rotation ratio from now on:

$$Re_S = \frac{u_S d}{\nu} = \frac{2}{1+\eta}\left|Re_1 - \eta Re_2\right|, \tag{2.23}$$

$$R_\Omega = \frac{2\Omega_{rf}d}{u_S} = (1-\eta)\frac{Re_1 + Re_2}{\left|Re_1 - \eta Re_2\right|} = \frac{(1-\eta)}{\eta}\frac{(\eta + \mu)}{\left|1 - \mu\right|} \tag{2.24}$$

2.3   Laminar Couette solution

If the flow inside the TC geometry is laminar, equation (2.13) can be solved analytically
[116]. Therefore, infinitely long cylinders, the absence of external forces, a stationary
and purely azimuthal flow ($\partial_t = 0$, $u_r = u_z = 0$, $u_\varphi = u_\varphi(r)$) and non-zero velocity
gradients only in the radial coordinate direction ($\partial_\varphi = \partial_z = 0$) are assumed. The axial
component of the Navier-Stokes equation vanishes and the radial component results in
the balance of the radial pressure gradient with the centrifugal force: $\rho u_\varphi^2/r = \partial_r p$. It
remains the azimuthal component, whose differential equation can be solved using the
Eulerian ansatz $u_\varphi(r) = r^n$. With $\omega = u_\varphi/r$, the laminar Couette solution takes the form

$$\omega_{lam}(r) = \frac{r_2^2\omega_2 - r_1^2\omega_1}{r_2^2 - r_1^2} - \frac{r_1^2 r_2^2(\omega_2 - \omega_1)}{r_2^2 - r_1^2}\frac{1}{r^2} = A_{lam} + \frac{B_{lam}}{r^2}, \tag{2.25}$$

with the boundary conditions $\omega(r_1) = \omega_1$ and $\omega(r_2) = \omega_2$. When the angular velocity
and the radial coordinate are normalized using the formulas $\tilde{\omega} = (\omega - \omega_2)/(\omega_1 - \omega_2)$
and $\tilde{r} = (r - r_1)/(r_2 - r_1)$, the laminar profile becomes independent of the cylinder
speeds and the radius ratio η is the family parameter:

$$\tilde{\omega}_{lam}(\tilde{r}) = \frac{1}{1 - \eta^2}\left[\left(\frac{\eta}{\tilde{r}(1 - \eta) + \eta}\right)^2 - \eta^2\right]. \tag{2.26}$$

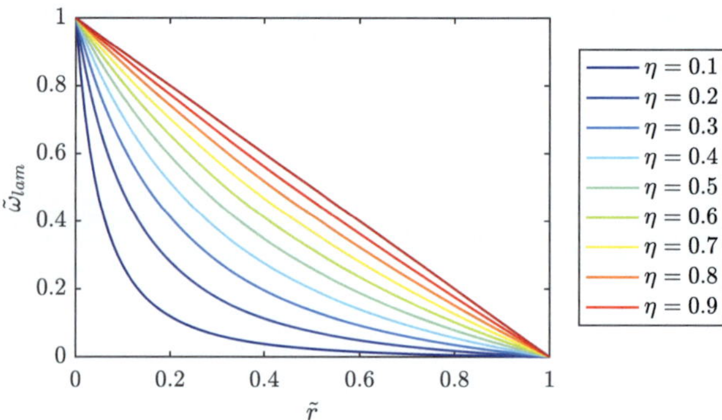

FIGURE 2.3: Normalized laminar Couette solution for various radius ratios η.

Note, that this normalization does not hold in case of $\omega_1 = \omega_2$, the so-called solid body
rotation. Figure 2.3 illustrates the physical meaning of the radius ratio as a measure for
the curvature of the cylinder walls affecting the flow. When $\eta \to 1$, the laminar Couette
flow in the TC system approaches to the plane Couette flow, where $\tilde{\omega}_{lam} \sim \tilde{r}$ and no
curvature exists. The corresponding boundary conditions of equation (2.26) are $\tilde{\omega}(\tilde{r}_1) = 1$
and $\tilde{\omega}(\tilde{r}_2) = 0$. Another important quantity concerning the laminar Couette flow is the
so-called neutral line $r_{n,lam}$. It is the radial location of the gap, where the azimuthal or
angular velocity component vanishes. With $\omega_{lam} = 0$, the neutral line is given by

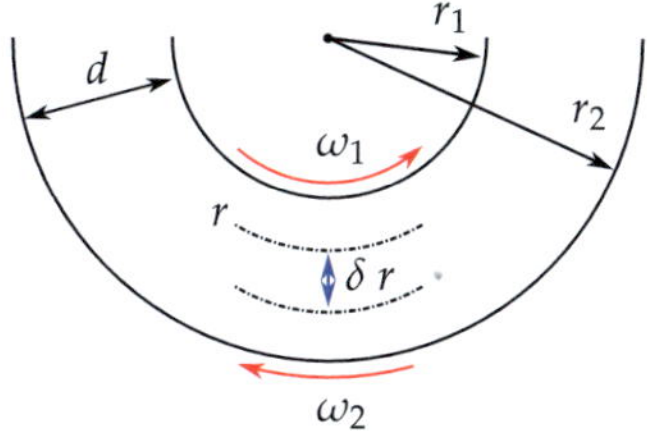

FIGURE 2.4: Sketch of two fluid rings in a TC flow as assumption for Rayleigh's stability criterion. Representation adopted from Prigent *et al.* [97].

$$r_{n,lam}(\mu) = r_1 \sqrt{\frac{1-\mu}{\eta^2 - \mu}}. \tag{2.27}$$

2.4 Centrifugal instability

In general, flows with curved streamlines like in a TC geometry can be centrifugally unstable. A classical tool to evaluate the stability is the linear stability analysis. There, the initial stable flow state is perturbed and the resulting equations of motion are linearized. If the perturbation abates again, the flow is called stable. However, here only two simple physical arguments are considered to decide about stability or instability, as this method gives more insight into the physics. The subsequent explanations follow closely the study of Esser and Grossmann [36].

2.4.1 Rayleigh's inviscid stability criterion

Within this section, the main ideas and findings of Rayleigh's stability analysis are recapitulated. One can assume an initially laminar Couette flow according to equation (2.25), an inviscid fluid ($\nu \nabla^2 u = 0$) and an axis-symmetric flow ($\partial_\varphi = 0$). Then, the azimuthal component of the Navier-Stokes equation (2.13) becomes

$$D_t \left(r u_\varphi\right) = D_t L = 0. \tag{2.28}$$

The angular momentum $L = r u_\varphi$ is conserved for all fluid particles. Now, two fluid rings are considered at the radii r and $r + \delta r$, respectively (see Figure 2.4). If these fluid rings swap their positions, their kinetic energy per mass $E_{kin} = u_\varphi^2/2 = L^2/(2r^2)$ has to decrease for stability. Accordingly, by subtracting the kinetic energy per mass of the fluid rings of the initial state from the one after the swap, the famous stability criterion from Rayleigh follows as

$$-\delta E_{kin} = \frac{1}{2} \left[\left(\frac{L(r+\delta r)}{r}\right)^2 + \left(\frac{L(r)}{r+\delta r}\right)^2 - \left(\frac{L(r)}{r}\right)^2 - \left(\frac{L(r+\delta r)}{r+\delta r}\right)^2 \right] \tag{2.29}$$

$$= \frac{\partial_r L^2(r)}{r^3} \left(\delta r\right)^2 > 0. \tag{2.30}$$

δr can be assumed as infinitesimally small, and if the angular momentum in equation (2.29) is replaced by the laminar Couette solution as initial flow state, the condition for stability takes the form

$$\mu > \eta^2. \tag{2.31}$$

2.4.2　Viscous stability criterion

If the idea of section 2.4.1 is extended to viscous flows according to Esser and Grossmann [36], the stability line for real fluids can be expressed analytically. The movement due to the swap of the two fluid rings causes dissipation of kinetic energy in the fluid. The friction energy loss per mass can be expressed as

$$\delta E_{vis} \sim \left(\frac{\nu}{l^2}\delta u_r\right)\delta r, \tag{2.32}$$

with the radial velocity of the swap δu_r and the typical length scale of the ring l. Instability sets in, when the kinetic energy exceeds the viscous damping on a timescale smaller than the viscous time $\tau_\nu = l^2/\nu$. Reformulating the radial velocity as $\delta u_r = \delta r/\delta t$ with $\delta t \sim l^2/\nu$, the stability boundary is defined by the relation $\delta E_{kin} \sim \delta E_{vis}$, leading to

$$-\frac{\partial_r L^2}{r^3}\left(\delta r\right)^2 \sim \frac{\nu^2}{l^4}\left(\delta r\right)^2. \tag{2.33}$$

Inserting equation (2.25) in the form of the laminar angular momentum in relation (2.33) yields

$$\left[Re_S\frac{\eta^2-\mu}{\eta|1-\mu|}\right]^2\left[\frac{r_{n,lam}^2}{r^2}-1\right] \sim \left(\frac{d}{l}\right)^4. \tag{2.34}$$

Up to know, the parameter l is still unknown and can be defined with the following assumptions. l represents the fraction of the gap, which becomes possibly unstable and can be modeled as $l = \alpha_{EG}d$, with the constant α_{EG} much smaller than 1 for co-rotating cylinders. In case of counter-rotation, a neutral surface appears in the gap and the effective gap width for instability reduces to $d_{n,lam} = r_{n,lam} - r_1$. In addition, the neutral line is some kind of soft border, which means that an instability can grow beyond this line to some extent. As measure for this extension, the parameter a is introduced. Accordingly, l can be defined as

$$l = \alpha_{EG}d\Delta_{EG}\left(a\frac{d_{n,lam}}{d}\right), \text{ with } \Delta_{EG}\left(x\right) = \begin{cases} x, & \text{if } x < 1 \\ 1, & \text{if } x \geq 1 \end{cases} \tag{2.35}$$

Finally, the radial position $r_{p,EG}$, where the motion of the fluid rings sets in, is defined as the gap center, because there the radial motion of the Taylor vortices is strongest. Including also counter-rotating flow states,

$$r_{p,EG} = r_1 + \frac{d}{2}\Delta_{EG}\left(a\frac{d_{n,lam}}{d}\right) \tag{2.36}$$

is chosen. In case of pure inner cylinder rotation ($\mu = 0$), equation (2.34) can be transformed, finding the critical shear Reynolds number $Re_{S,crit}(\mu = 0)$:

$$Re_{S,crit}(\mu = 0) = \frac{1}{\alpha_{EG}^2} \frac{1 + \eta}{\eta \sqrt{(1 - \eta)(3 + \eta)}}. \tag{2.37}$$

Here, the parameter α_{EG} contains all constant factors and can be fixed by matching the term $Re_{S,crit}(\mu = 0)$ to the value of the linear stability analysis for $\eta \to 1$, where this value is known very precisely. From Chandrasekhar [21], $\alpha_{EG} = 0.1556$ is found. Further, for co- and counter-rotating flow states, the stability line is defined as

$$\left(\frac{Re_{S,crit}}{Re_{S,crit}(\mu = 0)} \frac{\eta^2 - \mu}{\eta^2 |1 - \mu|} \right)^2 \frac{r_{n,lam}^2 - r_{p,EG}^2}{r_{p,EG}^2} \frac{(1 + \eta)^2}{(1 - \eta)(3 + \eta)} = \Delta_{EG} \left(a \frac{d_{n,lam}}{d} \right)^{-4}. \tag{2.38}$$

Further, the parameter a as measure for the extension of the instability beyond the neutral surface is given by

$$a(\eta) = (1 - \eta) \left[\sqrt{\frac{(1 + \eta)^3}{2(1 + 3\eta)}} - \eta \right]^{-1}, \tag{2.39}$$

from the condition, that the stability line is continuously differentiable.

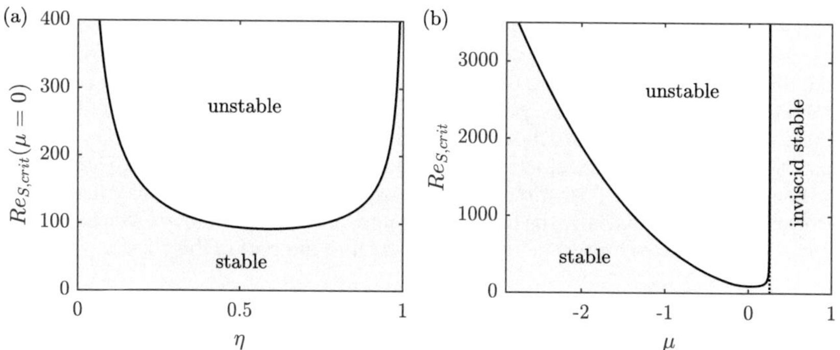

FIGURE 2.5: Diagram of primary centrifugal instability in TC flows. (a) Stability diagram for $\mu = 0$ as a function η. (b) Stability diagram for $\eta = 0.5$ as a function of Re_S and μ. The black dotted line indicates the Rayleigh stability line. Figure modified from Esser and Grossmann [36].

The stability diagrams related to the equations (2.29), (2.37) and (2.38) are shown in Figure 2.5. For pure inner cylinder rotation, a relatively low Reynolds number is needed in the case of medium gap TC flows to make the flow unstable, while for very small or large gaps, the critical shear Reynolds number increases drastically. In addition, the stability curve has a minimum around $\eta \approx 0.59$, indicating an asymmetry between both limiting cases. The stability curve as function of $Re_{S,crit}$ and μ for $\eta = 0.5$ demonstrates, that an increasing value of counter-rotation delays the transition to an unstable flow state. Further, the Rayleigh line and the stability line merge in the low co-rotating regime and viscosity can be seen as a damping force for the growth of the instability.

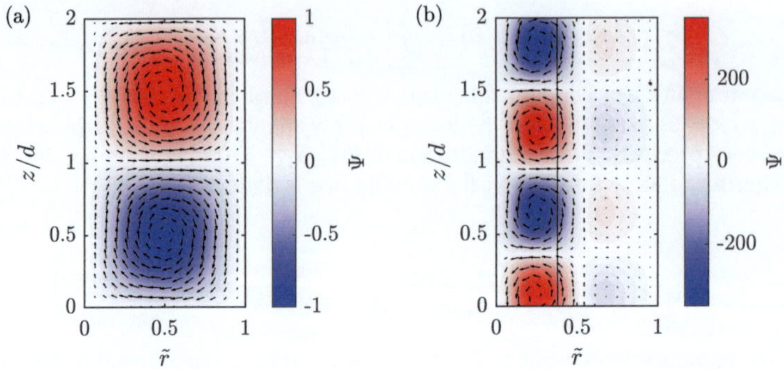

FIGURE 2.6: Streamfunction Ψ and velocity vectors in a meridional plane, when the centrifugal instability sets in, modified from Taylor [121]. (a) Pure inner cylinder rotation ($\mu = 0$) in the limiting case of $d \ll r_1$. Ψ is normalized to the interval ± 1. (b) Counter-rotation at $\mu = -1.5$ and $\eta = 0.94$. Black line indicates the neutral line from equation (2.27).

The appearance of the flow, when instability induced by symmetrical disturbances sets in, is depicted in Figure 2.6 according to Taylor [121]. For pure inner cylinder rotation, the primary centrifugal instability, namely Taylor vortex flow, consists of alternating vortices rotating in opposite direction. Each vortex fills the whole gap with an axial and radial dimension equal to the gap width d. For stronger counter-rotation, the centrifugal instability is limited to an inner part of the gap, which is again filled with alternating vortices of opposite direction. The vortices end at the radial location, where u_φ vanishes, which is however not identical to the predicted neutral line (black line in Figure 2.6(b)) based on equation (2.27). This growth of the instability beyond $r_{n,lam}$ was considered in the former stability analysis by the factor a, and now its validity is shown. The importance of these Taylor vortices for turbulent TC flows is one main part of this thesis.

2.5   Angular momentum transport and torque

This section is based on Eckhardt *et al.* [33] under the assumption of an incompressible fluid and infinitely long cylinders. Averaging the azimuthal component of the Navier-Stokes equation (2.13) over cylindrical surfaces, which are coaxial to the rotation axis, and time indicated by $\langle \cdot \rangle_{A_{TC}(r),t} = \langle \cdot \rangle_{\varphi,z,t}$, and eliminating the axial velocity component u_z with the continuity equation (2.10), one finds

$$0 = \partial_r \left(r^3 \left[\langle u_r \omega \rangle_{\varphi,z,t} - \nu \partial_r \langle \omega \rangle_{\varphi,z,t} \right] \right). \tag{2.40}$$

The quantity inside the round brackets is constant over the radius of the gap, as the radial gradient vanishes, and is also independent of φ, z and t due to the average. According to Eckhardt *et al.* [33], this quantity is called conserved transverse current of azimuthal motion or transport of angular momentum J_ω with

$$J_\omega = r^3 \left[\langle u_r \omega \rangle_{\varphi,z,t} - \nu \partial_r \langle \omega \rangle_{\varphi,z,t} \right]. \tag{2.41}$$

By inspecting equation (2.41), it becomes obvious that the relevant transport quantity in TC flows is the angular velocity. The first term reflects the convective transport and the second one the viscous part. Further, J_ω is directly proportional to the torque $\mathcal{T}$ that is necessary to apply at the cylinders to keep them at a constant speed:

$$J_\omega(r_{1,2}) = -r_{1,2}^3 \nu \partial_r \langle \omega \rangle_{\varphi,z,t}(r_{1,2}) = \frac{\mathcal{T}}{2\pi\ell\rho}. \tag{2.42}$$

This result is of great importance, as it illustrates that the overall angular momentum transport can be measured directly by the torque at the wall. In case of a laminar flow, the radial velocity component u_r vanishes and equation (2.25) can be inserted into equation (2.41):

$$J_\omega^{lam} = 2\nu r_1^2 r_2^2 \frac{\omega_1 - \omega_2}{r_2^2 - r_1^2}. \tag{2.43}$$

To non-dimensionalize J_ω, it is divided by its corresponding laminar value and the resulting quantity is called pseudo-Nusselt number in analogy to heat transport problems, as e.g. the Rayleigh-Bénard flow:

$$\mathrm{Nu}_\omega = \frac{J_\omega}{J_\omega^{lam}}. \tag{2.44}$$

Nu_ω measures the overall angular momentum transport in terms of the corresponding purely viscous transport. Therefore, $\mathrm{Nu}_\omega = 1$ holds for laminar flows. Further, the momentum transport is directly connected to the energy dissipation rate ϵ. By multiplying the three components of the Navier-Stokes equation (2.13) with the three velocity components u_r, u_φ and u_z, respectively, taking the sum, volume average the resulting equation and considering stationarity, ϵ is defined as

$$\epsilon = \frac{2}{r_2^2 - r_1^2} (\omega_1 - \omega_2) J_\omega, \tag{2.45}$$

$$\epsilon_{lam} = \frac{4\nu r_1^2 r_2^2}{(r_1 + r_2)^2} \left(\frac{\omega_1 - \omega_2}{d} \right)^2. \tag{2.46}$$

Similar to the approach for the momentum transport, only the increase of the energy dissipation rate due to convection compared to the laminar one is of interest. Therefore, the idea of the so-called wind is introduced. The wind is interpreted as the velocity field including a radial and axial component, which causes the difference between the laminar $(u_{r,lam} = u_{z,lam} = 0)$ and turbulent transport. So, a wind Reynolds number can be defined as

$$Re_W = \frac{U_W d}{\nu}, \tag{2.47}$$

with U_W as the typical amplitude of the wind, which can be calculated as standard deviation of the radial or axial velocity. Accordingly, the wind dissipation rate is

$$\epsilon_W \equiv \epsilon - \epsilon_{lam} = 4\nu \frac{r_1^2 r_2^2}{(r_1 + r_2)^2} \left(\frac{\omega_1 - \omega_2}{d} \right)^2 (\mathrm{Nu}_\omega - 1). \tag{2.48}$$

2.6   Analogy of Taylor-Couette, Rayleigh-Bénard and pipe flow

A fascinating fact about the TC flow is, that there are close analogies to the RB flow and
the pipe flow. Similar to the Taylor vortex flow evolving for pure inner cylinder rotation as supercritical transition in TC flow [121], Bénard rolls are formed in RB flow as
first instability, when a critical temperature difference is reached, with the same critical
point and wave number [97]. Further, the subcritical transition in a TC flow due to the
occurrence of turbulent spots or spirals is comparable to the one found in Poiseuille pipe
flow [38]. Both flows become turbulent if the lifetime of the localized turbulent regions
exceeds their decaying rate [4, 5, 12, 54].

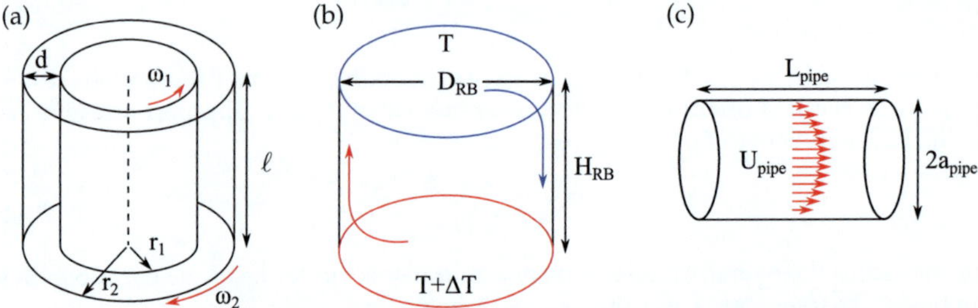

FIGURE 2.7: Comparative sketches of (a) the Taylor-Couette, (b) Rayleigh-Bénard and (c) pipe
geometry, modified from Eckhardt *et al.* [32].

By adequately averaging the equations of motion, similar expressions for the exact relations of the global transport quantities J and the dissipation rates ϵ_W can be found
in all three systems, as shown by Eckhardt *et al.* [32]. To reveal these analogies, equation (2.48) has to be transformed in case of TC flow. Therefore, a pseudo-Prandtl number
$\sigma_{Pr} = 2^{-4}(1 + \eta)^4 \eta^{-2}$ is introduced, which is a geometric parameter of the curvature,
and the Taylor number Ta, which is directly connected to the shear Reynolds number by
$Ta = \sigma_{Pr}^2 Re_S^2$. Now, the transport of angular momentum and the wind energy dissipation
rate in the TC flow are given by

$$J_\omega = r^3 \left[\langle u_r \omega \rangle_{A_{TC},t} - \nu \partial_r \langle \omega \rangle_{A_{TC},t} \right] , \tag{2.49}$$

$$\epsilon_W = \frac{\nu^3}{d^4 \sigma_{Pr}^2} Ta \left(\mathrm{Nu}_\omega - 1 \right) . \tag{2.50}$$

To recapitulate for TC flow, A_{TC} denotes cylindrical surfaces at constant r, which is used
to average the equations of motion. For the RB flow, the conserved heat flux J_T is given,
when the energy equation is averaged over planes A_{RB} parallel to the end plates at height
z, and the wind dissipation rate ϵ_W follows from an identical analysis to what has been
done in Section 2.5 in TC flow:

$$J_T = \langle u_z T \rangle_{A_{RB},t} - \kappa \partial_z \langle T \rangle_{A_{RB},t}, \tag{2.51}$$

$$\epsilon_W = \frac{\nu^3}{H_{RB}^4 Pr^2} Ra \left(\text{Nu}_T - 1 \right). \tag{2.52}$$

$Ra = \alpha_T g \Delta T H_{RB}^3 / (\kappa \nu)$ denotes the Rayleigh number, $Pr = \nu / \kappa$ the Prandtl number and $\text{Nu}_T = J_T / J_T^{lam}$ the heat flux Nusselt number, with the temperature T, the thermal expansion coefficient α_T, the gravitational acceleration g, the temperature difference ΔT, the cell height H_{RB} and the thermal diffusivity κ. Obviously, Ta and Ra, Nu_ω and Nu_T as well as σ_{Pr} and Pr are the analogous dimensionless numbers in both systems. While in TC, the angular velocity is convected by the radial velocity u_r, in RB the temperature T is convected by the axial velocity u_z. When the same analysis is done for pipe flow, using coaxial cylinders at constant r as surface A_{pipe} for the average, one finds

$$J_{u_z} = \frac{1}{r} \left[\langle u_r u_z \rangle_{A_{pipe},t} - \nu \partial_r \langle u_z \rangle_{A_{pipe},t} \right], \tag{2.53}$$

$$\epsilon_W = \frac{4 U_{pipe} \nu^2}{a_{pipe}^3} Re_{pipe} \left(\text{Nu}_{u_z} - 1 \right), \tag{2.54}$$

with the cross-sectional averaged velocity U_{pipe}, the pipe radius a_{pipe} and the Reynolds number $Re_{pipe} = 2 a_{pipe} U_{pipe} / \nu$. Here, J_{u_z} represents the axial velocity flux, where the axial velocity is convected by the radial velocity u_r. The prefactors r^3 in TC and r^{-1} in pipe for the conserved quantities J are a stringent consequence of the fact that they have to be constant across the named surfaces $A_{TC,RB,pipe}$. In summary, the analogy between all three systems for the exact relations of transport J and dissipation rate ϵ_W is significant and illustrates, that findings in TC flows can be possibly generalized to other shear flows and even thermal convective flows.

2.7 Main parameters and dimensionless numbers

Within this section, the main parameters and dimensionless numbers, which are used throughout this thesis, are summarized. The geometry of the TC system is defined by the gap width d, the radius ratio η and the aspect ratio Γ:

$$d = r_2 - r_1, \qquad \eta = \frac{r_1}{r_2}, \qquad \Gamma = \frac{\ell}{d}. \tag{2.55}$$

To describe the driving of the system in terms of shear and rotation, the shear Reynolds number Re_S and the ratio of angular velocities μ are used:

$$Re_S = \frac{2}{1+\eta} |Re_1 - \eta Re_2| = \frac{2 r_1 r_2 d}{(r_1 + r_2)\nu} |\omega_2 - \omega_1|, \tag{2.56}$$

$$\mu = \frac{\omega_2}{\omega_1}. \tag{2.57}$$

The global response of the flow to changes in the driving is quantified by the pseudo-Nusselt number Nu_ω and the wind Reynolds number Re_W:

$$\mathrm{Nu}_\omega = \frac{J_\omega}{J_\omega^{lam}} = \frac{r_1 r_2}{2\pi\ell\rho\nu^2 d^2}\mathcal{T}Re_S, \tag{2.58}$$

$$Re_W = \frac{U_W d}{\nu} = \frac{\sigma(u_r)d}{\nu}. \tag{2.59}$$

$\sigma(u_r)$ represents the standard deviation of the radial velocity component.

Chapter 3

Current state of research

Within this section, the main findings in current TC research in the turbulent regime are presented. The focus is set to the angular momentum transport (or torque) and the underlying flow organization, which leads to the following topics: (*i*) What are the effective scaling laws relating momentum transport and driving, expressed non-dimensionally as a Reynolds number (Re_1 or Re_S), (*ii*) how does the momentum transport depend on the rotation ratio μ, (*iii*) what are the characteristics of the velocity field, (*iv*) how do the large-scale structures known as Taylor vortices influence the transport, (*v*) to what extent do these large-scale structures interact with small-scale plumes, and (*vi*) what role plays the curvature. Additional information concerning TC research can be found in various reviews, e.g. about TC history [30], hydrodynamic instabilities [26, 39] or fully turbulent TC flows [51]. This Chapter follows closely the introductions of the papers reported in Chapter 1 [41, 43, 45].

Torque scaling estimates and results for pure inner cylinder rotation

Typically, the effective scaling of the momentum transport is analyzed expressed in terms of the dimensionless torque $G = \mathcal{T}/(2\pi\ell\rho\nu^2)$ for pure inner cylinder rotation ($\mu = 0$) by assuming a power-law ansatz of the form $G \sim Re_{1,S}^{\alpha}$, which translates for the pseudo-Nusselt number to $\mathrm{Nu}_\omega \sim Re_{1,S}^{\alpha-1}$. Besides, to emphasize the similarities between the TC flow and RB convection, also the effective scaling of Nu_ω with the Taylor number Ta is used, leading to $\mathrm{Nu}_\omega \sim Ta^{\frac{\alpha}{2}-\frac{1}{2}}$. By using marginal stability analysis, King *et al.* [65] and Marcus [72] predicted effective scalings for the torque in the spirit of what had been done for RB flow by Malkus and Veronis [69]. Marginal stability theory is based on a separation of the flow into an inner boundary layer, a bulk flow, and an outer boundary layer. Both BLs are assumed to be laminar and the bulk flow as well as the BLs are marginally stable, according to Rayleigh's stability criterion. As a consequence, the radial profile of the angular momentum $L = \omega r^2$ has to be flat in the bulk flow, which is in good agreement with the findings of previous studies [17, 28, 111]. Matching of the L-profiles at the separation borders and independence of the momentum transport on the radial coordinate leads to a predicted scaling exponent of $\alpha = 5/3$, identical to the one for RB flow and the calculations of Barcilon and Brindley [8]. Another prediction for the effective torque scaling was derived by Lathrop *et al.* [66] using a Kolmogorov type argument. They assumed that the energy dissipation rate ϵ is constant in the inertial range and has no length scale dependence. This condition translates into the limit of an infinite Reynolds number, where viscous effects can be neglected. In contrast to Lathrop, here the shear velocity u_S is used as the characteristic velocity difference and the gap width d as the characteristic length, leading to the relation

$$\epsilon = \frac{2}{r_2^2 - r_1^2}\,(\omega_1 - \omega_2)\,\nu^2 G \sim \frac{u_S^3}{d^3} \quad \longleftrightarrow \quad G \sim \frac{\eta}{(1-\eta)^2}Re_S^2, \tag{3.1}$$

with $\alpha = 2$. The same exponent was also found by Doering and Constantin [27], directly derived from the Navier-Stokes equation as an upper bound. Further, Eckhardt *et al.* [33] deduced a prediction for the effective torque scaling in analogy to the RB flow by distinguishing the contributions of the bulk and the BL to the overall transport. For not-too-wide gaps ($\sigma_{Pr} \approx 1$) and laminar boundary layers, they found a scaling exponent of $\alpha = 5/3$, when the boundary layers are dominant and $\alpha = 2$, when the bulk flow is dominant.

Contrary to these predictions, no pure power-law scaling was detected within numerous numerical and experimental studies. Wendt [138] performed direct torque measurements for three radius ratios ($\eta = 0.68$, 0.85, and 0.935), finding $\alpha = 1.5$ in the interval $4 \times 10^2 \leq Re_1 \leq 10^4$ and $\alpha = 1.7$ for $Re_1 > 10^4$. Instead of a partially constant exponent, Lathrop *et al.* [66] experimentally identified a monotonically increasing exponent from $\alpha = 1.23 - 1.87$ in the range of $8 \times 10^2 \leq Re_1 \leq 1.2 \times 10^6$, and a sharp discontinuity in the α-Re_1-curve around $Re_1 = 1.3 \times 10^4$ for $\eta = 0.724$. These measurements have been repeated by Lewis and Swinney [68] with one order of magnitude higher accuracy, confirming the previous results. A similar radius ratio of $\eta = 0.716$ has been investigated by van Gils *et al.* [130, 132] and Huisman *et al.* [60], where a nearly constant scaling exponent of $\alpha = 1.78 \pm 0.06$ was identified for Reynolds numbers beyond the aforementioned change in the scaling. In addition, Merbold *et al.* [78] measured the torque for $\eta = 0.5$, where α was increasing up to $Re_S = 6 \times 10^5$, before it settled to a nearly constant value around $\alpha = 1.65 \pm 0.03$. Again, a transitional behavior of the exponent was detected around $Re_S \approx 8 \times 10^4$–$10^5$. Similar results for a wide range of radius ratios of $\eta = 0.714$, 0.769, 0.833 and 0.909 with a universal scaling exponent of $\alpha = 1.78$ were found by Ostilla-Mónico *et al.* [87] for Taylor numbers larger than $Ta = \sigma_{Pr}^2 Re_S^2 > 10^{10}$. The only investigation of the momentum transport for even wider gap TC flows at a radius ratios of $\eta = 0.35$ were performed by Burin *et al.* [19], finding $\alpha = 1.6 \pm 0.1$ for $2 \times 10^3 \leq Re_1 \leq 10^4$ and $\alpha = 1.77 \pm 0.07$ for $2 \times 10^4 \leq Re_1 \leq 2 \times 10^5$. There, J_ω was measured only locally and indirect by LDV. It has to be stressed that the last mentioned result contradicts the previous investigations, where a monotonic increase of the transitional Reynolds number with decreasing η were reported. In the study of Burin *et al.* [19] however, the bulk dominated regime ($\alpha > 5/3$) is reached at $Re_1 \approx 2 \times 10^4$ much earlier than in the case of $\eta = 0.5$. Besides, in all presented investigations a transitional behavior was found in the scaling exponent, whose origin was linked by Ostilla-Mónico *et al.* [86] to a shear instability in the BLs, where laminar BLs would transition to turbulent ones. This picture unifies the aforementioned scaling predictions. At low Reynolds numbers, the momentum transport is limited by the laminar boundary layers ($\alpha = 5/3$), while at large Reynolds numbers, the featureless turbulent bulk is the limiting factor ($\alpha = 2$). In analogy to the RB flow, the regime, where the BLs are laminar is called '*classical*', and the regime with turbulent BLs is called '*ultimate*'.

Torque dependence on the amount of rotation

When independently rotating cylinders come into play, the torque exhibits an additional dependency on the rotation ratio μ and it can be shown, that a separation ansatz of the form $G = f_1(\mu) f_2(Re)$ is valid [31, 91]. For medium and wide-gap TC flows, the torque depicts a broad maximum in the slight counter-rotating regime when the shear Reynolds number is kept constant and only the rotation rate μ is changed. The maximum location strongly depends on the radius ratio and was found in various numerical and experimental studies. Paoletti and Lathrop [90] discovered within their direct torque measurements for $\eta = 0.724$ a maximum at $\mu_{\max} = -0.33$, which was confirmed later by van Gils *et al.* [130] for a similar radius ratio of $\eta = 0.716$ at $\mu_{\max} = -0.33 \pm 0.04$. Moreover, Merbold *et al.* [78] could show, again by use of direct torque measurements, that the maximum location shifts for wider gaps ($\eta = 0.5$) to less strong counter-rotating rates around $\mu_{\max} = -0.2 \pm 0.02$. Both identified maximum locations have been verified numerically by Brauckmann and Eckhardt [15], where the torque maximizing rotation rates were located at $\mu_{\max} = -0.195 \pm 0.019$ for $\eta = 0.5$ and at $\mu_{\max} = -0.361 \pm 0.054$ for $\eta = 0.71$. Further confirmation was given by numerical simulations of Ostilla-Mónico *et al.* [87], finding $\mu_{max}(\eta = 0.5) = -0.22$ and $\mu_{max}(\eta = 0.714) = -0.33$. It is worth mentioning that the location of the torque maximum slightly changes with the forcing and only reaches a fixed value at large enough Re_S. Further, studies for radius ratios of $\eta > 0.8$ are excluded due to the fact, that the flow behavior changes, as described by Brauckmann *et al.* [17]. There, instead of one '*broad maximum*', one '*narrow maximum*' connected to strongly correlated vortices and a '*broad maximum*' due to a streamwise invariant strong vortical flow have been detected.

A first attempt to explain the physical mechanism behind the single torque maximum for medium and wide-gap TC flows was made by van Gils *et al.* [130] with the so-called angle bisector hypothesis. van Gils *et al.* [130] argued that the most unstable point in the parameter space is the location equally distant from both Rayleigh stability lines $\mu = \eta^2$ and $\mu = \infty$, which defines the rotation ratio of the torque maximum μ_b by:

$$\mu_b = \frac{-\eta}{\tan\left[\dfrac{\pi}{2} - \dfrac{1}{2}\arctan\left(\eta^{-1}\right)\right]}. \tag{3.2}$$

While the predicted rotation ratio of the torque maximum becomes $\mu_b = 0.368$ for the radius ratio $\eta = 0.716$ in good agreement with the measurements of van Gils *et al.* [130] of $\mu_{\max} = -0.33$, it deviates noticeable for $\eta = 0.5$, where it becomes $\mu_b = -0.309$ in contrast to the measurements of Merbold *et al.* [78] with $\mu_{\max} = -0.198$.

Another prediction, linking the location of the torque maximum to the onset of intermittency in the gap and a strengthening of large-scale Taylor vortices, was developed by Brauckmann and Eckhardt [15]. Their theory states that turbulent Taylor vortices in the slight counter-rotating regime can extend the theoretical neutral line by the factor $a(\eta)$, which is the so-called parameter of vortex extension according to equation (2.39). $a(\eta)$ depends only slightly on η and reaches values in the range of $1.4 - 1.6$ [36]. If this extension exactly ends at the outer cylinder wall, the Taylor vortices are most pronounced and cause a maximum in transport. For even higher counter-rotating rates, the vortices are restricted to an inner gap region with a laminarized outer region. As the momentum transport has to be constant over the whole gap, intermittent bursts appear, which are flushing from the unstable inner gap region into the stable outer gap region. Based on the stability calculations of Esser and Grossmann [36], the prediction of Brauckmann and Eckhardt [15] can be written as:

$$\mu_p(\eta) = -\eta^2 \frac{(a^2(\eta) - 2a(\eta) + 1)\eta + a^2(\eta) - 1}{(2a(\eta) - 1)\eta + 1}. \tag{3.3}$$

A connection between the torque maximum and the onset of intermittency was also identified via LDV measurements by van Gils *et al.* [130], and the strengthening of Taylor vortices was shown by Ostilla-Mónico *et al.* [86]. These studies are discussed in more detail in the next section. In comparison, the prediction yields $\mu_p(\eta = 0.5) = -0.191$ and $\mu_p(\eta = 0.71) = -0.344$ in very good agreement with the aforementioned findings. This reflects that Taylor vortices play a prominent role in the momentum transport in the fully turbulent regime for the '*broad maximum*' identified by Brauckmann *et al.* [17].

Characteristics of the mean velocity field

Based on the angular momentum transport described in the previous section, specific characteristics of turbulent TC flows need to be discussed. The most simple comparison of different flow states can be done based on the radial profiles of the angular or azimuthal velocity, averaged over cylindrical surfaces. As described by King *et al.* [65] and Marcus [72], the profile can be separated into two boundary layers and a bulk flow. At low Reynolds numbers before the transition in the torque scaling exponent occurs, the BLs are laminar with a shape according to the Prandtl-Blasius theory [11, 95]. If the Reynolds number is sufficiently large, the BLs become turbulent due to a shear instability, and the profiles depict a pronounced logarithmic region, as predicted by van Kármán [102], which is also valid in other shear flows. Such logarithmic boundary layers in TC flows have been found and analyzed for instance by Huisman *et al.* [56] or Grossmann *et al.* [52]. On the other hand, the bulk flow, whose laminar shape has been introduced in equation (2.25), becomes turbulent at a much lower Reynolds number, resulting in a flat profile. This flatness reflects the effective mixing due to turbulence. As the momentum transport is also a measure for the mixing efficiency, the slopes of the profiles have to change with μ. Ostilla-Mónico *et al.* [87] reported that the smallest gradient of angular velocity coincides with the rotation ratio of the torque maximum for $\eta = 0.5$, 0.714, 0.833 and 0.909, while Brauckmann *et al.* [17] could validate these results only for $\eta \geq 0.8$. The difference may be caused by differences in the investigated Re_S-regime and in the evaluation methods, and measurements at Re_S far beyond those of Brauckmann *et al.* [17] ($Re_S = 2 \times 10^4$) could give more clarity. Further, the radial location of the neutral line is of special interest for the prediction of Brauckmann and Eckhardt [15]. Ostilla *et al.* [82] found that the detachment of the neutral line from the outer cylinder wall starts at pure inner cylinder rotation for $Re_S \leq 4620$, while van Gils *et al.* [130] identified a coincidence of μ_{max} with the neutral line detachment for $Re_S > 9 \times 10^5$. Both studies were carried out at a radius ratio of $\eta = 0.71$. Clearly, the discrepancy in the results comes again from Re_S differences and the latter finding is what fits with the prediction of Brauckmann and Eckhardt [15]. However, the LDV measurements of van Gils *et al.* [130] only have been performed at one cylinder height, while the flow in the presence of Taylor vortices strongly depends on the axial coordinate.

For stronger counter-rotation rates ($\mu < \mu_{max}$), a radial partitioning exists in the flow by the neutral line. While the inner gap region is turbulent and centrifugal unstable, the outer gap region is stabilized by the outer cylinder rotation [21], featuring penetrating intermittent turbulent bursts [15]. These bursts have been experimentally observed by Coles [23] and Andereck *et al.* [3], and numerically investigated by Coughlin and Marcus [25]. In the latter study, the bursts are described as *'temporal oscillation between a spatially laminar flow and turbulence'*. Later, van Gils *et al.* [130] could verify the existence of such bursts by LDV measurements in a TC flow at Reynolds numbers of $Re = \mathcal{O}(10^6)$. By evaluating the PDFs of the angular velocity for different radial positions, they found a unimodal distribution in the inner turbulent region and a bimodal distribution in the outer region, caused by intermittent bursts. Brauckmann *et al.* [17] summarized, that the radial flow partitioning results in differences between the inner and outer gap region, concerning their linear stability, intermittency and fluctuations.

Next to the azimuthal velocity component, the TC flow field consists of a non-zero radial as well as axial velocity component within the turbulent regime. The global strength of this secondary flow can be measured in terms of the wind Reynolds number according to equation (2.47), which can be evaluated based on the standard deviation of the radial or axial velocity component:

$$Re_W = \frac{\sigma(u_{r,z})d}{\nu} \sim Re_S^\beta \tag{3.4}$$

Similar to the pseudo-Nusselt number, a change in the scaling exponent β of Re_W with the shear Reynolds number is predicted by Grossmann and Lohse [49, 50], based on the TC-RB analogy. The exponent β in the classical regime is predicted as $\beta = 6/7$ and was measured by van der Veen *et al.* [128] as $\beta = 0.842$ for $\eta = 0.5$ and $\mu = 0$, in very good agreement. Further, Huisman *et al.* [60] found an exponent of $\beta = 0.99$, close to the predicted one of $\beta = 1$ in the ultimate regime for $\eta = 0.716$ and $\mu = 0$. Both studies used the standard deviation of the radial velocity component to calculate Re_W. The transition from the classical to the ultimate regime as well as the dependence of Re_W on μ, however, has not yet been measured in TC flows.

Large-scale turbulent Taylor vortices

It has been shown by Huisman *et al.* [57], that Taylor vortices can survive in the fully turbulent regime up to very high Reynolds numbers of $Re = \mathcal{O}(10^6)$, and the maximum in torque has been linked by Brauckmann and Eckhardt [15] to a formation and strengthening of turbulent Taylor vortices. Therefore, these large-scale structures have been in focus of various studies concerning their contribution to the overall momentum transport. Performing numerical simulations for pure inner cylinder rotation with varying axial length of laminar and turbulent Taylor vortices, Brauckmann and Eckhardt [14] could show, that there exists an optimal wavelength $\lambda_{TV} = (\Gamma d)/n_{TV}$ of $\lambda_{TV} = 0.84d$ and $\lambda_{TV} = 0.97d$, respectively, where the pseudo-Nusselt number is maximal. Further, Martínez-Arias *et al.* [74] found out experimentally for $\eta = 0.909$, that a smaller wavelength of turbulent Taylor vortices, and therefore a higher total number of vortices across the gap, result in a larger transport of angular momentum in the classical regime and vice versa in the ultimate regime. In contrast, Huisman *et al.* [57] reported for $\eta = 0.716$ a higher Nusselt number for a larger number of vortices in the ultimate regime for differential rotation, measuring the torque only at a central segment of the inner cylinder. They could further show, that multiple stable vortex states exist in the fully turbulent regime. This finding was verified and extended to a much larger parameter space by van der Veen et al. [127]. By decomposing the flow field into its turbulent and large-scale contributions and calculating the associated Nusselt number fractions, Brauckmann and Eckhardt [15] and Brauckmann et al. [17] could demonstrate the importance of large-scale Taylor rolls to the overall transport. They discovered smooth changeovers from a turbulence-dominated transport for co- and high counter-rotation to a large-scale-circulation-dominated transport in the torque maximum region for different η at $Re_S = 2 \times 10^4$.

Another interesting phenomenon regarding Taylor vortices is the reappearance of azimuthal waves in the turbulent Taylor vortex regime. Walden and Donelly [135] measured the point-wise radial velocity component close to the OC boundary layer for $\mu = 0$, $\eta = 0.875$ and different aspect ratios. They found for $28 \leq Re/Re_{crit} \leq 36$ a regime of reappearance for $\Gamma \geq 25$, based on sharp peaks in the power spectra. Here, Re_{crit} is the critical Reynolds number for the onset of Taylor vortex flow. Later, Takeda [118] acquired time-resolved axial profiles of the axial velocity component using an ultrasonic measurement technique for $\eta = 0.904$ and $\Gamma = 20$. The azimuthal waves were identified based on a Fourier- and POD-analysis in the range of $23 \leq Re/Re_{crit} \leq 36$. Their results show good agreement with those of Walden and Donelly [135]. In another study, Wang et al. [136] performed planar PIV measurements in a meridional plane for $\eta = 0.733$ and $\Gamma = 34$. They captured the reappearance of azimuthal waves for $20 \leq Re/Re_{crit} \leq 38$ based on spatial correlations. Note that the three aforementioned studies are all based on pure inner cylinder rotation ($\mu = 0$). More recently, Merbold et al. [76] performed flow visualizations in TC flow for $\eta = 0.5$ and $Re_S = 5 \times 10^3$. They found an axial oscillation of the turbulent Taylor vortices in the range of $\mu \in [-0.15, -0.3]$, which includes the rotation ratio of the torque maximum ($\mu_{max} = -0.2$). In summary, the large-scale turbulent Taylor rolls seem to feature an instability mechanism similar to the one in the laminar regime, which, however, has not yet been detected in highly turbulent TC flows.

Interaction of turbulent Taylor vortices and small-scale plumes

Ostilla et al. [82] showed in analogy to RB flow for $\eta = 0.714$, that the large-scale rolls consist and are driven by small-scale unmixed plumes. They calculated for $\mu = 0$ and $\eta = 0.714$ the radial profiles of the angular velocity ω at specific axial positions of the large-scale Taylor vortices; namely at the vortex inflow, vortex center and vortex outflow. The vortex inflow is characterized by the ejection of plumes from the OC in conjunction with a mean radial velocity that points away from the OC. In contrast, the outflow features plume ejections from the IC with a mean radial velocity component directed from the IC to the OC. In between the in- and outflow, the radial velocity component becomes zero in the middle of the gap, which denotes the location of the vortex center (see Figure 3.1). In regions, where plumes are ejected from the cylinder walls, i.e. in the vortex inflow at the outer wall and in the vortex outflow at the inner wall, the radial profiles of ω have a more pronounced logarithmic shape in the corresponding boundary layer [88]. It is worth mentioning that in the ultimate regime, the profiles become logarithmic also in the absence of dominant large-scale rolls [56]. For $\eta = 0.5$ in the classical regime, van der Veen *et al.* [128] calculated the velocity of such plumes based on planar PIV measurements, performed at different heights, which further confirmed the connection between small-scale plumes and large-scale vortices.

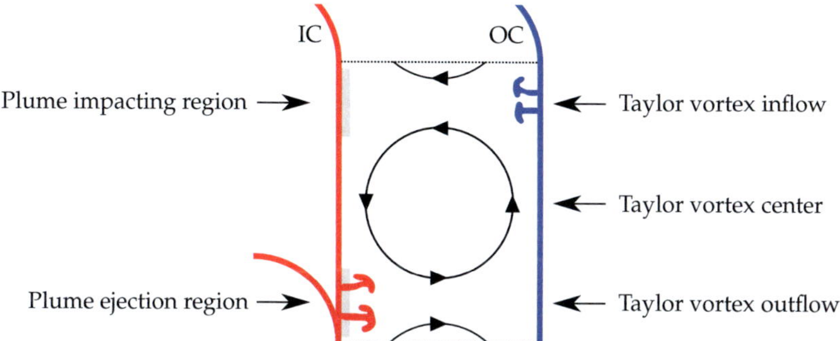

FIGURE 3.1: Sketch of a Taylor vortex, consisting of unmixed plumes. The axial locations of the Taylor vortex inflow, center and outflow, as well as plume ejection and impacting regions are shown.

Within the context of both TC and RB flow, many studies in the literature focused on establishing a distinct connection between the transport of angular momentum (or heat in RB flow) and the structures inherent to these flows, both for large-scale rolls and small-scale plumes. A comparative study of the PDFs of the Nusselt number in TC and RB, calculated over cylindrical surfaces and horizontal planes respectively, has been performed by Brauckmann *et al.* [16]. As a reference point for the comparison, they chose $Re_S = 2 \times 10^4$ for TC and $Ra = 10^7$ for RB, where the Nusselt numbers are identical for both flows. They found that the PDFs of the net transport have the same asymmetric shape with differences in the width of the tails in the BL regions. These differences can be attributed to different shapes and detachment frequencies of the plumes. Moreover, the PDFs of the angular momentum and temperature fluctuations depict a cusp-like shape with pronounced exponential tails as a consequence of the appearance of intermittent bursting plumes. Similar analyses of heat flux PDFs in RB flow have been performed by Shishkina and Wagner [110]. They found that the instantaneous heat flux fluctuates around zero and not around the volume-averaged Nusselt number, along with a broadening of the tails with increasing Ra. Shang *et al.* [109] revealed that the asymmetry of the PDFs of Nu_T, which mainly occurs in the tails, arises from correlated temperature and velocity signals produced by thermal plumes. These plumes lead to large but rare positive events of heat flux. However, the results of Shang *et al.* [109] are only based on point-wise measurements. For TC flow, the statistics of Nu_ω were analyzed for $\mu = 0$ by Huisman *et al.* [58] without any connection to specific flow structures.

Another approach investigating structures in TC flow is to analyze the kinetic energy spectra. Dong [28] performed DNSs for $\eta = 0.5$, $Re = 8 \times 10^3$ and the OC at rest. Strikingly, he found a peak at small scales in the axial spectra of the radial velocity component. According to Dong [28], the underlying structures can be specified as herringbone streaks. The DNSs of Ostilla-Mónico *et al.* [88] in the boundary layer regions revealed a peak in the azimuthal and axial spectra of the radial velocity component at large wavenumbers, which also indicates the existence of small-scale structures (named here plumes). Their simulations were done for $\eta = 0.909$ and $\mu = 0$ at $Re_S \geq 10^5$. It has to be stressed that the energy spectra in TC flow do not follow the classical Kolmogorov scaling for homogeneous and isotropic turbulence of -5/3 [28, 58, 68, 88, 133].

Curvature effects

According to Dubrulle *et al.* [31], the curvature number R_C, which occurs in equation 2.21 when an advection shear term is introduced, is the appropriate parameter to quantify the curvature forces in TC flows:

$$R_C = \frac{d}{\sqrt{r_1 r_2}} = \frac{1-\eta}{\sqrt{\eta}} \tag{3.5}$$

R_C is defined as the reciprocal of the geometrical mean of both cylinder radii normalized by the gap width and is, therefore, a function of the radius ratio η. It can be understood as a measure for the importance of the additional coupling between wall-normal (radial) and streamwise (azimuthal) velocities induced by the curvilinear coordinates. This coupling causes strong modifications in shear flows. When studying boundary layers over curved surfaces, Bradshaw [13] noted '*the surprisingly large effect exerted on shear-flow turbulence by curvature of the streamlines in the plane of the mean shear*'. In Figure 3.2, R_C is depicted as a function of η. As η approaches to 1, R_C becomes zero and curvature effects vanish. In this situation, TC flow becomes rotating plane Couette flow [17]. Further, the condition $R_C \to 0$ enables a better comparison of TC and RB flows based on their analogies [32], as curvature effects are absent in RB flows.

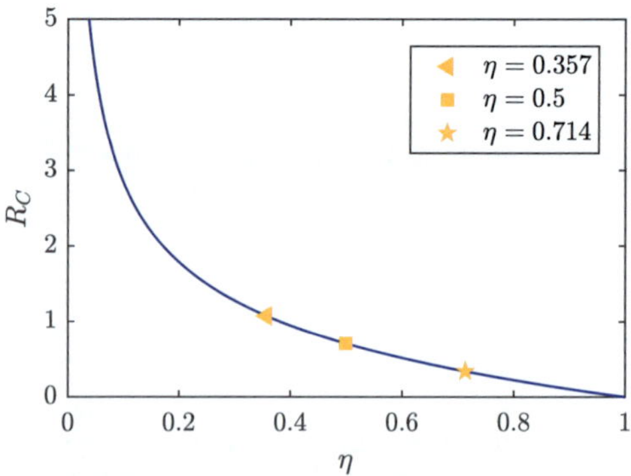

FIGURE 3.2: Curvature number R_C as function of the radius ratio η. Symbolds represent the radius ratios used within this thesis.

Although the rotation ratio of the torque maximum (μ_{max}) strongly depends on η and therefore on the curvature of the system, the curves of momentum trransport as a function of the rotation number R_Ω collapse for a wide range of η and R_Ω, when an appropriate compensation is used. Dubrulle *et al.* [31] noted, that the dimensionless torque G, normalized by the value for pure inner cylinder rotation $G(\mu = 0)$, depends only little on η. In the same manner, Paoletti *et al.* [91] related G to $G(R_\Omega = 0)$, which also results in a collapse of momentum transport for many radius ratios. Recently, Brauckmann *et al.* [17] performed numerical simulations for radius ratios between 0.5 and 1, where they compensated the Nusselt number by $Re_S^{0.78}$. They found a collapse of the Nu_ω-R_Ω-curves for all R_Ω for $\eta \geq 0.9$, while the collapse is limited for $\eta = 0.8$ to $R_\Omega \geq 0.1$ and for $\eta = 0.71$ and $\eta = 0.5$ to $R_\Omega \geq 0.25$. The borders for the agreement of the curves are related to the existence of the already mentioned radial partitioning of the flow. Therefore, they concluded, that the appearance of radial flow partitioning introduces a strong dependence of the Nusselt number on the radius ratio or rather curvature. This dependence vanishes either for large enough R_Ω or $\eta \geq 0.9$.

Chapter 4

Experimental setup

The Taylor-Couette flow has been investigated for over 100 years, starting with Couette [24] and Mallock [70], who built up concentric cylinder experiments to determine the viscosity of fluids. Henceforward, many different experimental facilities were designed to address specific scientific questions. The aim of this thesis is to investigate fully turbulent Taylor-Couette flows for different radius ratios, which leads to several requirements for the setup: (*i*) Fully turbulent flow states must be accessible by selecting an appropriate combination of fluid and cylinder speed; (*ii*) the cylinders must be able to rotate independently to realize co- and counter-rotating conditions, and (*iii*) the radius ratio should be variable; (*iv*) optical access to the flow is required to use optical measurement techniques to analyze the velocity field within the gap; (*v*), the measurement of the torque should be possible to directly quantify the angular momentum transport.

	Delft	Maryland	Twente		Cottbus	Princeton	Göttingen
optical access	OC	OC	OC EPW	OC TP	OC	OC	OC
torque	√	√	√	√	√	×	×
η	0.92	0.73	0.72 0.77 0.83 0.91	0.5 0.71	0.5	0.35	0.71 0.88 0.98

TABLE 4.1: Actual turbulent TC facilities. Characteristics are chosen as optical accessibility, a direct torque measurement unit and the radius ratio. Abbreviations are OC for outer cylinder, EPW for end plate window and TP for top plate. City names represent the following papers: Delft [99], Maryland [66, 68, 90], Twente [59, 131], Cottbus [78], Princeton [105, 106] and Göttingen [4].

If the accessibility of the fully turbulent regime and independently rotating cylinders are presumed, a few TC facilities already exist and have been used in the past. In the early days, Wendt [138] build up a TC experiment for three radius ratios ($\eta = 0.68$, 0.85 and 0.935) and measured the torque at the inner cylinder wall as well as velocity profiles using Pitot tubes. The disadvantage of his design is the lack of an optical access, which is why it is not mentioned in table 4.1. Further, Schartmann et al. [105, 106] constructed a short annulus TC apparatus (referred to as Princeton) with variable end plate configurations to study the quasi-Keplerian regime, which is of special interest for astrophysical problems. In Maryland, a TC experiment is used in the very high Reynolds number regime ($Re_1 \sim 10^6$) to study the global angular momentum transport, local shear stresses and velocity fluctuations [66, 68, 90]. Ravelet *et al.* [99] also measured the torque and the velocities in a TC facility in Delft for a very narrow gap. Two outstanding TC facilities have been build up at the University of Twente, the so-called Twente turbulent Taylor-Couette facility (T^{3}C, van Gils *et al.* [131]) and the boiling Twente Taylor-Couette facility (BTTC, Huisman *et al.* [59]). The T^{3}C was designed to investigate single phase as well as bubbly TC flows at high Reynolds numbers and different radius ratios regarding the momentum transport, and the BTTC can operate with boiling liquids. Furthermore, at the BTU Cottbus-Senftenberg, a wide-gap TC apparatus has been constructed for precise torque measurements at very high cylinder speeds [78]. Finally, a new TC experiment has been build up in Göttingen for different radius ratios to investigate subcritical transitions and the effects of boundary conditions [4]. As regards the test facilities mentioned above, table 4.1 lists their main features with regard to the requirements of the TC apparatus used for this thesis. The comparison points out that most of the available facilities focus on narrow and very narrow gaps. Accordingly, the investigation of wide-gap TC flows is highly needed to complete the parameter space concerning the radius ratio. In addition, the optical accessibility is usually given only by the outer cylinder, while transparent end plates are rare. This current state of experiments led to the design of the TC facility used within this thesis.

4.1 The top-view Taylor-Couette Cottbus (TvTCC) facility

The TvTCC experiment has already been outlined roughly in [44, 45, 76, 128]. Here, all technical details and the overall design are presented. The basic concept is a simple structure, in order to quickly mount and dismount the facility. This requirement is important, as different inner cylinders and working fluids are used. Further, the seals and bearings, which are integrated into the system, are chosen to minimize the friction of the inner cylinder, as the torque is measured at its wall. A CAD model and a technical drawing of the experiment are depicted in Figure 4.1. The base of the experiment is an aluminum footplate, which contains two drill holes, one with a thread to connect a drain plug in order to fill the gap and one for a temperature sensor (thermocouple type K, -50 up to $300°C$, $\pm1.5\%$), measuring the fluid's temperature close to the bottom plate. Further, it contains two concentric grooves for the outer cylinder and its O-ring seal, and a central turning slot for the bottom bearing of the inner cylinder. The acrylic and therefore transparent outer cylinder holds two glued flange rings, which are displaced relative to its end in order to fit into the footplate groove. The form-fitted connection between the flange ring and the plate is realized by 12 screws. In addition, a bottom plate is screwed on top of the footplate to enclose the gap. On the upper side of the experiment, the acrylic top plate with its mounted bearing seat is connected to the outer cylinder in the same way as on its lower side, leading to end plates rotating together with the outer cylinder. The bearing seat volume is linked to the gap volume, resulting in a buffer volume to ensure that the whole gap is filled without remaining air bubbles. In addition, it contains two bleeder holes for the filling and emptying process. Inside the outer cylinder, the inner cylinder configuration is placed. It contains a stainless steel inner shaft, which is seated by two stainless steel single-row groove ball bearings in the footplate and the upper bearing seat above the top plate, to rotate relative to the outer cylinder. This configuration is the lower limit of the accessible radius ratio. To set up a smaller gap, the inner shaft can be replaced by a thicker shaft, or different inner aluminum anodized cylinders can be mounted on the inner shaft. In the latter case, aluminum fittings are screwed on both ends into the inner cylinder, which contain O-rings to seal against the inner cylinder and a rotary shaft seal plus stainless steel trantorques (*Fenner Drives*), to connect and seal it against the inner shaft. Therefore, the inner cylinder is free of fluid in its interior. Note that the use of different inner cylinders requires different top and bottom plates.

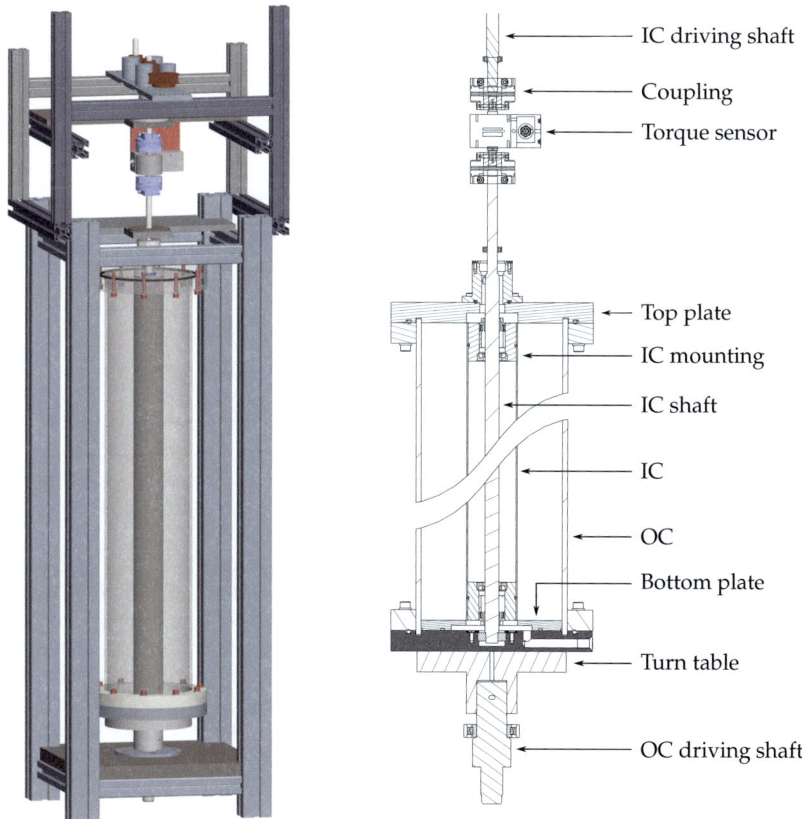

FIGURE 4.1: Illustration of the top-view Taylor-Couette Cottbus experiment. (a) CAD image of the experiment with its stand and inner cylinder driving. (b) Technical drawing of a central cut through the experiment excluding the stand and cylinder driving units.

Above the upper bearing seat, the inner shaft is seasoned against the stand, again by a stainless steel single-row groove ball bearing. The end of the inner shaft is connected to a shaft-to-shaft rotary torque sensor (*Lorentz Messtechnik GmbH*, DR2500, 2 Nm, 0.1%) via a torsionally stiff coupling. On the other side of the sensor, its shaft is linked to the IC driving shaft by use of an identical coupling. In a similar way, the footplate is placed on a turntable, which is driven by the OC driving shaft. This shaft is seated against the stand. Both cylinders are driven by 200 W *Maxon* DC motors via a belt drive. Especially on the top side, this belt drive is connected to the IC driving shaft using four pulleys to prevent shear forces acting on the inner power train. In table 4.2, the different geometrical configurations, achievable cylinder speeds, and Reynolds numbers are summarized.

Geometrical parameter						
ℓ	[mm]			700		
r_1	[mm]	7	14	25	35	50
r_2	[mm]			70		
$d = r_2 - r_1$	[mm]	63	56	45	35	20
$\eta = r_1/r_2$	[−]	0.10	0.20	0.36	0.50	0.71
$\Gamma = \ell/d$	[−]	11.1	12.5	15.6	20	35
Maximum cylinder speeds						
n_1	[rpm]			2200		
n_2	[rpm]			500		
Maximum Reynolds numbers ($\nu = 10^{-6}\,\mathrm{m^2/s}$)						
Re_1	[−]	1.02×10^5	1.81×10^5	2.59×10^5	2.82×10^5	2.30×10^5
Re_2	[−]	2.31×10^5	2.05×10^5	1.65×10^5	1.28×10^5	7.33×10^4
$Re_S(\mu = 0)$	[−]	1.85×10^5	3.01×10^5	3.82×10^5	3.76×10^5	2.69×10^5

TABLE 4.2: Design parameters of the top-view Taylor-Couette Cottbus experiment, including geometry, cylinder speeds and maximum Reynolds numbers at room temperature; with cylinder length ℓ, inner and outer radius $r_{1,2}$, gap width d, radius ratio η and aspect ratio Γ.

To conclude, the TvTCC experiment enables a simple implementation of different inner cylinders (different η) and full optical access to the flow through the OC as well as the top plate. Further, a torque sensor is directly integrated into the power train of the inner shaft, leading to a torque signal which is superimposed by friction losses of bearings and seals. To minimize this friction, only three low friction single-row groove ball bearings are integrated into the inner shaft bearing and rotary shaft seals have been waived. At last, the experiment is not temperature-controlled but -monitored, which leads during operation to a small but continuously increase of the fluid temperature due to friction losses.

4.1.1   Rotation and temperature control

The TvTCC experiment is controlled by the software program LabVIEW®. To adjust the speeds of the inner and outer cylinder, a PID-controller is applied, which uses as input data the actual cylinder velocities, measured by light barriers at a frequency of 10 Hz. It is very important that the cylinder speeds are set accurately with only slight temporal fluctuations, as quasi-stationary measurements shall be performed. The accuracy of the cylinder speed control is illustrated in Figure 4.2(a). All data points for a constant μ are on a line crossing the origin. This indicates that the set and actual cylinder speeds match very well with a relative error of the mean smaller than 0.6%, regardless of the absolute

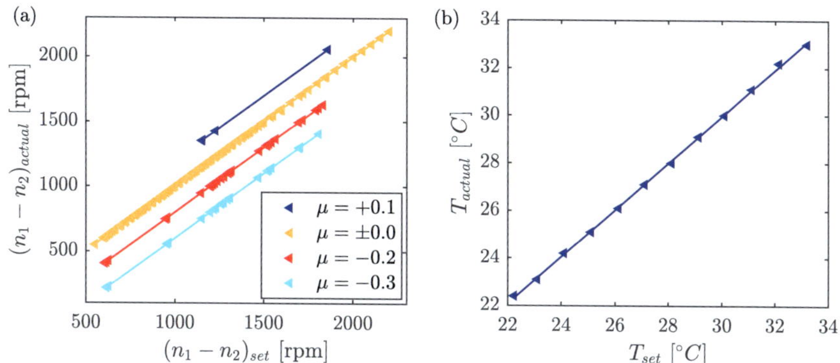

FIGURE 4.2: (a) Accuracy of cylinder rotation for different μ. Lines represent linear regressions. Error bars are in the size of the symbols. The relative error of the mean is smaller than 0.6 % and the standard deviation is smaller than 1 %. Data points are shifted for $\mu \neq 0$ for better visibility. (b) Calibration of the temperature sensor. The deviation of the set and the actual temperature is smaller than 0.8 %.

speed. For better visibility, the data points for $\mu \neq 0$ are shifted. Further, the local velocity fluctuations do not exceed a value of 1 %. Therefore, the TvTCC experiment is appropriate to adjust and investigate stable flow states. In addition to the cylinder speeds, the fluid temperature must also be taken into account, since the fluid properties such as density ρ or kinematic viscosity ν, and thus the shear Reynolds number Re_S are temperature dependent. Within the TvTCC experiment, the fluid temperature cannot be controlled but monitored with a thermocouple type K sensor (-50 up to $300°C$, $\pm 1.5\%$), which is inserted into the lower end of the gap through the footplate. Since fully turbulent flows are investigated, the mixing in the gap is very effective, and it can be assumed that the fluid temperature throughout the gap is distributed fairly homogeneously. The sensor setup is limited to sequential, point-wise measurement when the outer cylinder is at rest. Thus, it has to be ensured, that the temperature rise of the fluid during a measurement point is only slight. If this is the case, the temperature-dependent fluid properties can be corrected for each measurement point. As a first step, the temperature sensor is calibrated using a Julabo F25 Waterbath. As shown in Figure 4.2(b), the implemented temperature sensor measures the actual fluid temperature very accurately.

In order to study the temperature rise inside the gap due to friction losses further, quasi-stationary temperature measurements were performed for pure inner cylinder rotation with increasing inner cylinder speed using the high-viscosity silicone oils M5 and M10, the properties of which will be further described in the next section. The measuring protocol was standardized as follows: Each measurement point consists of a waiting time of 120 s, where the cylinder speeds are adjusted and the flow has time to settle. Thereafter, the fluid temperature T is measured, before the cylinder speed n_1 is recorded for another 90 s. The results for the two fluids at different initial temperatures are shown in Figure 4.3. As expected, the fluid temperature increases due to energy dissipation while the experiment is running. However, the increase in temperature is low when using high-viscosity fluids with fast cylinder speeds, resulting in an increase of only $1.4°C$ over 45.5 min in

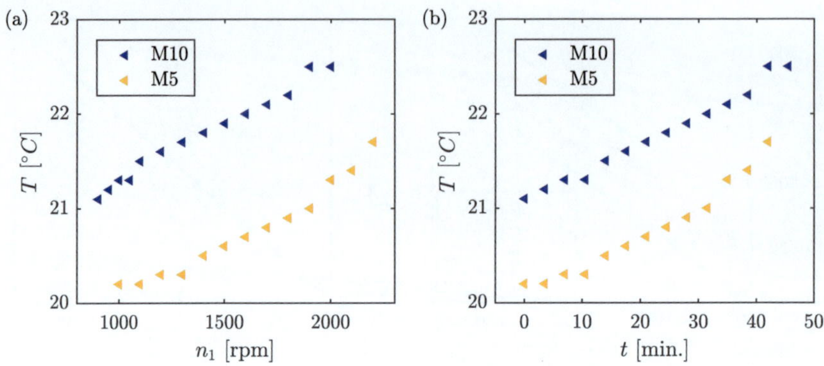

FIGURE 4.3: Exemplified temperature increase in the TvTCC experiment due to friction for $\mu = 0$. (a) Temperature increase as function of the inner cylinder speed. (b) Corresponding temperature increase as function of the experiment running time t.

the case of M10 and 1.5°C over 42 min at a lower initial temperature in the case of M5. From this, it can be concluded that it is sufficient to measure the temperature just at the beginning of a measurement point.

4.1.2   Working fluids

For the experiments performed within this thesis, distilled water and silicone oils of different kinematic viscosities ν are used to adapt the fluid properties to the particular measurement task.

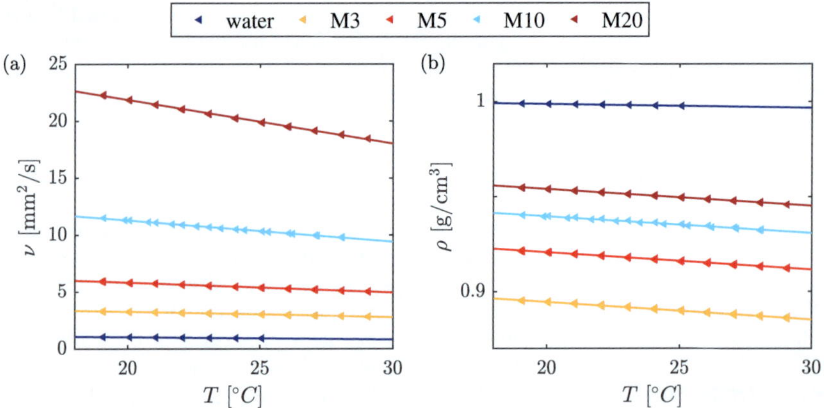

FIGURE 4.4: Viscosimeter measurements of the working fluids distilled water, M3, M5, M10 and M20. Solid lines represent linear fits for the data points. (a) Kinematic viscosity ν as a function of the temperature T. (b) Density ρ as a function of the temperature T.

While the maximal cylinder rotation is fixed by the presented driving, higher shear Reynolds numbers can be reached by reducing ν. Moreover, when the angular momentum transport is measured at low Reynolds numbers, a more viscous fluid enhances the torque signal. To provide accurate measurements, it is important to know the temperature dependence of the properties of the working fluids precisely, as the TvTCC experiment is not temperature controlled. Therefore, a Stabinger viscosimeter with integrated density measurement cell (Anton Paar, SVM 3000/G2, $\triangle\nu = \pm0.35\,\%$, $\triangle\rho = \pm0.0005\,\mathrm{g/cm^3}$) is used to measure the fluid's density ρ and the kinematic viscosity ν as a function of the temperature T. As shown in Figure 4.4, the kinematic viscosity and the density of all fluids studied decrease linearly with increasing temperature. The more viscous silicone oils also have a higher density, while distilled water has the lowest viscosity and highest density of all liquids considered. The linear relationship between the properties of the fluid and the temperature is given in table 4.3. The temperature dependence of the fluids with respect to ν is two orders of magnitude higher than for ρ and even increases with more viscous fluids.

	kinematic viscosity $\nu\,[\mathrm{mm^2/s}]$ [a]			density $\rho\,[\mathrm{g/cm^3}]$ [b]		
	a	b	@ 25°C	c	d	@ 25°C
distilled water	-0.022	1.480	0.919	-2.175×10^{-4}	1.003	0.998
M3	-0.046	4.175	3.023	-9.356×10^{-4}	0.913	0.890
M5	-0.088	7.612	5.402	-9.247×10^{-4}	0.939	0.916
M10	-0.189	15.05	10.34	-8.952×10^{-4}	0.958	0.935
M20	-0.385	29.57	19.95	-8.963×10^{-4}	0.972	0.950

TABLE 4.3: Linear fits for the kinematic viscosity ν and the density ρ of the fluids used within this thesis. The temperature is given in °C and the correlation values are always larger than 0.99.

[a]$\nu(T\,[°\mathrm{C}]) = aT + b$
[b]$\rho(T\,[°\mathrm{C}]) = cT + d$

4.2 The boiling Twente Taylor-Couette (BTTC) facility

In addition to experiments in the TvTCC apparatus, measurements were also carried out in BTTC facility in the framework of an EuHIT project, entitled '*Formation of large-scale vortices in the torque maximum region in Taylor-Couette flow (LSCTC)*'. The facility has been described by Huisman *et al.* [59] in all details. Here, only the main design parameters of the apparatus are mentioned (see table 4.4). According to its name, the BTTC was developed to study boiling liquids, requiring accurate temperature control of the inner and outer cylinders and the end plates, respectively. Therefore, the top plate and the OC contain cooling channels, which limits the optical access to the flow. Accordingly, another combination of OC and top plate was used for the investigations within this thesis. The BTTC consists of an inner cylinder made of copper, which is chrome-plated against corrosion, and a transparent outer cylinder made of acrylic glass. Both cylinders can rotate independently and confine the measurement gap. The inner cylinder is further separated into three segments with gaps of 2 mm in between. Both end plates rotate together with

Geometrical parameter	ℓ [mm]	r_1 [mm]	r_2 [mm]	d [mm]	η	Γ
	549	75	105	30	0.71	18.3
maximum cylinder speeds	n_1 [rpm]			n_2 [rpm]		
	± 1200			± 600		
maximum Reynolds numbers	Re_1		Re_2		$Re_S(\mu = 0)$	
($\nu = 10^{-6}\mathrm{m}^2/\mathrm{s}$)	2.8×10^5		2.0×10^5		3.3×10^5	

TABLE 4.4: Design parameters of the boiling Twente Taylor-Couette experiment, including geometry, cylinder speeds and maximum Reynolds numbers at room temperature; with cylinder length ℓ, inner and outer radius $r_{1,2}$, gap width d, radius ratio η and aspect ratio Γ [59].

the outer cylinder and the top plate is again made of transparent acrylic glass. The inner cylinder is tempered by a cooling circuit and equipped with three PT100 sensors at the heights $z = 0.22\ell$, $z = 0.49\ell$ and $z = 0.76\ell$. Both cylinders are driven by servo motors of 2.57 kW and 1.39 kW respectively, and magnet angle encoders are used to measure the angle of the shafts. By differentiating these angles, the rotation rates are recorded. In addition, a hollow flanged reaction torque transducer measures the torque over the whole length of the inner cylinder, including friction losses of bearings and seals. The experiment is designed to work with distilled water and refrigerants as FC-3248 or FC-87. For this study, distilled water was used. The operation of the experiment is realized by a LabVIEW® program.

Chapter 5

Measurement techniques

5.1 Flow visualization

The flow visualization technique is a frequently-used tool in fluid mechanics to analyze the topology of a flow qualitatively, allowing deep insights into the underlying physical processes [80, 125]. As most liquids and gases are optically transparent, their motion cannot be observed directly, and special techniques have to be used to make the flow visible. While Reynolds [101] showed in his famous pipe experiment the transition process from laminar to turbulent flows using dye, Prandtl [94] demonstrated the formation of vortices in liquid flows using small particles, and Mach visualized supersonic flow phenomena based on optical techniques like interferometry [100]. Good overviews of different visualization techniques are given by Merzkirch [79] and van Dyke [129]. This thesis is focused on liquid flows that are seeded with particles. Such a particle-based method has been widely used in TC flows to visualize e.g. the plurality of supercritical instabilities [3, 23, 121], the occurrence and lifetime of subcritical turbulent spots and spirals [4, 12, 18], or the variability of the number of turbulent Taylor vortices [74]. In these studies, combinations of working fluids and tracer particles were used as distilled water mixed with Kalliroscope or silicone oils mixed with aluminum flakes, excluding Taylor's work.

5.1.1 Setup and image processing

Flows have been visualized in the TvTCC experiment for a radius ratio of $\eta = 0.357$ and the silicone oil M3. The setup (see Figure 5.1(a)) consists of two halogen lamps mounted on the experiment stand at the height of about the top and bottom plate, respectively, and centered to the rotation axis. Both lamps are tilted into the direction of mid-height of the experiment to illuminate an axially middle segment of the flow. Further, an *Optronis* CR 3000x2 high-speed camera (1690×1710 px) sitting on a tripod was positioned in front of the experiment, capturing approximately 6 gap widths in the axial direction and the whole outer cylinder diameter in the radial direction, centered slightly above mid-height. To make the fluid motion visible, the silicone oil is suspended with anisotropic aluminum flake particles, which have an elliptic shape and an upper diameter of approximately $5\,\mu$m. As shown experimentally by Abcha *et al.* [1] using Kalliroscope flakes in a TC system, *'small anisotropic particles align with the flow streamlines by giving the precision on the velocity component which bears these alignment.'* In their setup, they used a light sheet visualisation in the radial-axial plane, which led to an intensity image of the reflected light that corresponds to the magnitude of the radial velocity component in good agreement with numerical simulations of Gauthier *et al.* [47]. For the present setup, where the flow is visualised in the azimuthal-axial plane, the intensity image should provide information on the azimuthal u_φ or axial velocity component u_z, meaning that bright areas are related to large absolute values of u_φ or u_z. Moreover, blue pigments (*BASF*, Heligen blue, K6850) have been added into the fluid to reduce its transparency and focus on the flow in the outer gap region. This is necessary, as a fully turbulent flow in a wide-gap configuration is visualized. When the experiment is not running, the aluminum and pigment particles settle down slowly. After approximately three days, all particles have been mixed out and gather at the bottom plate. However, a uniformity of suspension can be restored within a few minutes by setting up a fully turbulent flow state.

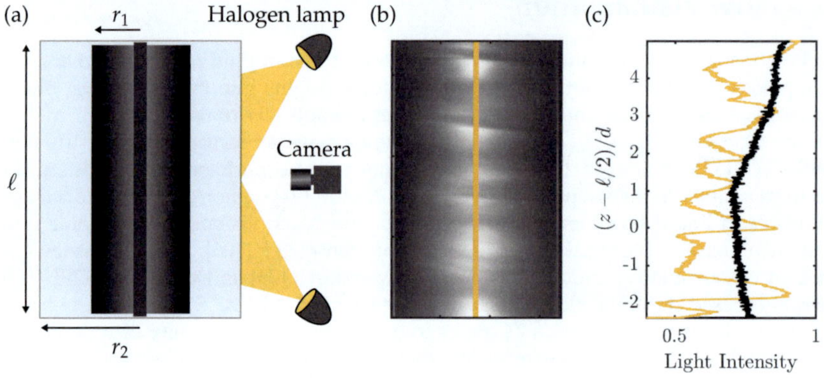

FIGURE 5.1: (a) Sketch of TC apparatus with visualization setup. (b) Time-averaged image over 2900 snapshots of the flow, recorded at 60 Hz for $Re_S = 3 \times 10^3$ and $\mu = 0$. Yellow line indicates the position of the evaluated axial light intensity profiles. (c) Axial light intensity profile ($\mathcal{I}$, yellow) along the line of (b). Black line indicates the axial light intensity profile for a laminar flow ($\mathcal{I}_{lam}$).

Figure 5.1(b) depicts for $Re_S = 3 \times 10^3$ and $\mu = 0$ an averaged flow image of 2900 snapshots, recorded at 60 Hz. Horizontal and equally spaced lines are visible in the image, which suggests, that Taylor vortices exist inside the gap, changing the amount of light reflected by the particles relative to a purely laminar flow. In Figure 5.1(c), the axial light intensity profile $\mathcal{I}$ along the yellow line of Figure 5.1(b) is shown with the same color. Aside, the black line corresponds to the axial intensity profile $\mathcal{I}_{lam}$, captured for a purely azimuthal flow. Its deviation from a vertical line is due to inhomogeneous illumination, which can be used to correct the intensity value of the recorded images by subtracting both signals: $\tilde{\mathcal{I}} = \mathcal{I} - \mathcal{I}_{lam}$. To reveal the flow structure more clearly, the corrected instantaneous light intensity profile of the depicted central axial line in Figure 5.1(b) is plotted against the time coordinate. This representation is called space-time diagram and shown in Figure 5.2. The gap is filled with 4 turbulent Taylor vortex pairs along an axial length of approximately 6 gap widths. The outflow regions with the up- and downwelling fluid appear like plaits, while the inflow regions look like straight lines, oriented into the time direction. Due to the existence of strong turbulent fluctuations, the whole image is interspersed with alternating bright and dark regions. Obviously, this visualization method is appropriate to capture large-scale turbulent structures like turbulent Taylor vortices.

FIGURE 5.2: Space-time diagram of a TC flow at $Re_S = 3 \times 10^3$ and $\mu = 0$, based on the light intensity $\tilde{\mathcal{I}}$ over a central, axial line and time. Axes normalization is done by the gap width d and the viscous time $\tau_\nu = d^2/\nu$. The figure is corrected for inhomogeneous illumination.

5.2 Torque measurements

The torque measurement technique is a widespread tool in rotating fluid mechanics research. Before over 200 years, it was primarily used to quantify the resistance of rotating machinery [137]. Motivated by the aim to determine the viscosity of fluids, this technique has been transferred to the flow confined by concentric cylinders. Mallock [70] built up a TC apparatus with a uniformly rotating outer cylinder and a stationary inner one, suspended from a single torsion wire. By measuring the deflection angle of the wire, he could calculate the torque at the inner cylinder. In later experiments, Mallock [71] used the same technique to measure the torque at the outer cylinder while rotating the inner one. Nearly at the same time, Couette [24] performed similar experiments rotating the outer cylinder, while measuring the torque at the inner one, again by a single torsion wire. However, he connected the wire via pulleys to a weighing pan. By adding weights into the pan during an experimental run until the IC was brought back into its initial position, he could directly measure the torque. A different technique to analyze the torque at the inner cylinder was used by Wendt [138]. He applied a pair of two inner cylinders, connected by a calibrated spring with an initial load. An extension or compression of the spring was transferred into a shift of a pencil that was confined by two electric contacts. Each of these contacts had its own current circuit with a lamp inside. When the rotation of the inner or outer cylinder was changed, the spring tension was adjusted by a spindle nut until the pencil did not touch one of these contacts anymore, indicated by the extinction of the lamps. Then, the spring tension is a measure for the friction torque at the inner cylinder. The outstanding innovation of this apparatus is the measurement of the torque, which is purely induced by the fluid without superposition of any bearing friction, and the possibility of measurements, while both cylinders rotate independently. Three years later, Taylor [120] performed also torque measurements to quantify the viscosity of fluids for either pure inner or outer cylinder rotation, using a torque thread similar to Couette and Mallock. The wire was connected to a scale pan, and by adding weights to it, he determined the weight needed for the start of co-rotation and the start of counter-rotation, respectively. The average of both values was used as a measure for the torque. Further, he repeated these measurements with air instead of a working fluid, to determine the torque induced by bearing friction and subtracted this value from the original signal. Nowadays, measurements of the torque in a TC flow are still of interest to analyses the angular momentum transport. The up-to-date facilities mostly have a similar configuration. The inner cylinder is separated into three segments with small gaps in between, and the torque is sensed only in the middle section by load cells or strain gauges. The separation of the IC is done to reduce the effect of end walls onto the torque signal. As the sensor is mounted in a rotating frame, the signal is transduced to a computer via a slip ring, and the calibration of the sensors is performed using standard weights hanging at specific radii on the IC middle segment. [66, 78, 131].

5.2.1 Integration of the torque sensor unit

The integration of a torque sensor into the TvTCC experiment has been illustrated in Section 4.1, and here the most important issues are summarized. To implement a sensor, that can run with different inner cylinder configurations, a simple arrangement of torque sensor inside the power train is used. The sensor is a contactless shaft-to-shaft rotary torque sensor (DR2500/M220-G21) of the company *Lorentz Messtechnik GmbH* with a nominal torque of ± 2 Nm and an accuracy of 0.1 %. It can be used up to rotation rates of 3×10^4 rpm, including the whole range of accessible inner cylinder rotation rates. The sensor features no internal bearing to prevent corruption of the signal and is directly connected to the measurement computer and integrated into LabVIEW® by a USB port. Further, the sensor is mounted vertically concerning its shaft orientation and fixed to the stand of the experiment. To connect the sensor's shaft with the shafts of the power train, torsionally stiff couplings (966 G50) of the same company with a maximum transmissible torque of 6.17 Nm and a speed limit of 4.4×10^4 rpm are used. In contrast to the previous mentioned up-to-date torque sensor units, here the torque is measured over the whole length of the inner cylinder, including end wall effects, and the torque signal is superimposed on the friction of three stainless steel single-row groove ball bearings. Therefore, each measurement point is measured twice in terms of the rotation rate of the cylinders, once with a fluid-filled gap and once with air. Afterward, both signals are subtracted to get the torque signal, which is induced by pure friction of the fluid. Note, that the general calibration of the sensor has been performed and certificated by the manufacturer.

5.2.2 Validation of the torque sensor unit

Within this section, the characteristics of the torque signal in the TvTCC experiment are discussed and its averaged values are compared with other studies. Therefore, a radius ratio of $\eta = 0.5$ is chosen, where different experimental and numerical results already exist. The subsequent validation measurements are performed quasi-stationary and the torque is analyzed for 180 s at a sampling frequency of 10 Hz. It is worth mentioning that also test measurements using an acquiring frequency well above the rotation frequencies of the cylinders have been done, to analyze the Fourier spectrum of the signal. There, no distinct peak appeared, which indicates, that the torque signal is not significantly corrupted by vibrations of the rotating experiment. This result is important, as the sensor is integrated directly into the power train, is mounted on the stand of the experiment

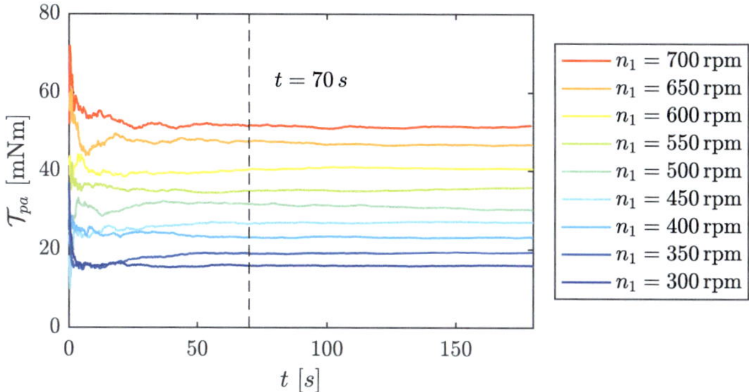

FIGURE 5.3: Progressing average of the torque $\mathcal{T}_{pa}$ as a function of the measuring time t for $\eta = 0.5$, $\mu = 0$ and different inner cylinder speeds. After 70 s, the average approaches to a constant value.

instead of a fixed location and features no internal bearing. To evaluate the required measuring time, that is needed for the average of the torque to be independent of the length of the signal, the progressing average of the torque $\mathcal{T}_{pa}$ is analyzed. With the number of samples n, $\mathcal{T}_{pa}$ is defined by

$$\mathcal{T}_{pa,n} = \frac{1}{n} \sum_{i=t_1}^{t_n} \mathcal{T}_i. \tag{5.1}$$

Figure 5.3 depicts the progressing average of the torque as a function of the measuring time t for pure inner cylinder rotation and 9 different cylinder speeds. Despite initial fluctuations, the progressing average approaches to a constant value after approximately 70 s and the final value increases with the IC rotation speed. Therefore, the measurement time can be reduced to e.g. 90 s. Further, the remaining relative measurement uncertainty of the torque average for all data presented within this thesis is always smaller than 5 %.

Next, the torque for $\mu = 0$ at different inner cylinder speeds and the fluids M3, distilled water and air is depicted in Figure 5.4(a). Again, the torque increases with the inner cylinder speed and at the same rotation rate, a more viscous liquid leads to a higher torque value. In the case of air as working fluid, the torque, which is assumed to be dominated by the friction forces of the bearings, is much smaller than the corresponding ones with the liquids. Note, that measurement points have been rejected, where the torque value using air exceeds 20 % of the value with the liquids. The torque difference of liquid and air in terms of the Nusselt number as a function of the shear Reynolds number is shown in Figure 5.4(b) in logarithmic scale. The Nusselt number increases with the shear Reynolds number nearly exponentially and due to the use of two liquids with different viscosities, a shear Reynolds number range of $2.6 \times 10^4 \leq Re_S \leq 2.7 \times 10^5$ is accessible. As comparison, direct torque measurement data of Merbold *et al.* [78] and numerical data of Ostilla-Mónico *et al.* [86, 87] are added for the same radius ratio. Especially at lower shear Reynolds numbers, the agreement with both studies is very good. In the upper Re_S-regime, no numerical data exist and a noticeable but acceptable discrepancy to the data of Merbold *et al.* [78] is found.

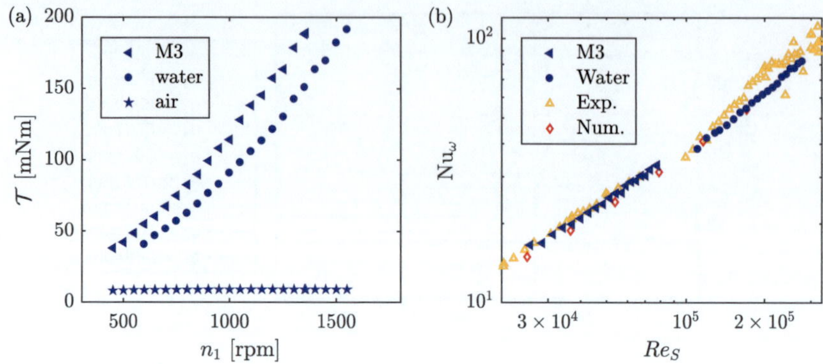

FIGURE 5.4: Validation of the torque sensor unit for $\eta = 0.5$ and $\mu = 0$. (a) Torque $\mathcal{T}$ as a function of n_1 using air, M3 and distilled water as working fluids. (b) Nusselt number as a function of the shear Reynolds number. The torque data of air are subtracted from the one with the liquids. Measurements are compared to experiments of Merbold *et al.* [78] (Exp.) and numerical simulations of Ostilla-Mónico *et al.* [86, 87] (Num.) at the same η.

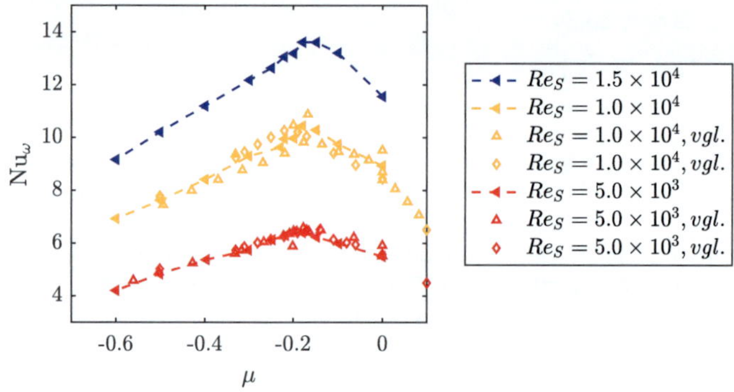

FIGURE 5.5: Nusselt number Nu_ω as a function of the rotation ratio μ for three different shear Reynolds numbers Re_S and $\eta = 0.5$. Results of this study for the two lower Reynolds numbers (filled triangles) are compared to direct numerical simulations (open diamonds) and direct torque measurements (open upward triangles) of Merbold *et al.* [78].

In Figure 5.5 torque measurements of this study are compared to measurements and numerics of Merbold *et al.* [78], where the shear Reynolds number is kept constant and only the rotation ratio μ is changed. A nearly perfect match with the data of Merbold *et al.* [78] is found with a relative deviation smaller than 5 %. In addition, a common maximum of the Nusselt number appears around $\mu = -0.2$. It is worth mentioning that in the study of Merbold *et al.* [78], the torque was measured only at a middle segment of the inner cylinder and the end plates were kept stationary, while the numerical simulations of Ostilla-Mónico *et al.* [86, 87] and Merbold *et al.* [78] featured periodic boundary conditions and therefore no end wall effects. In the present setup, the torque is measured over the whole length of the inner cylinder and the end plates are fixed to the outer cylinder. Taking these differences into account, the excellent match of the torque data demonstrates that end wall effects play a negligible role in fully turbulent Taylor-Couette flows, and the current torque measurement unit provides an accurate analysis of the angular momentum transport.

5.3 Particle image velocimetry (PIV)

PIV is an optical measurement technique to quantify the displacement of small tracer particles conveyed by a fluid [80, 125]. In contrast to single-point measurement techniques like LDV or hot-wire anemometry, PIV is capable of capturing the instantaneous velocity field of up to 10^5 grid points simultaneously, giving access to the investigation of large-scale flow patterns like coherent structures [141]. This fact explains the great importance of PIV especially for the investigation of turbulent flows. The development and application of PIV to fluid flows started in 1977 based on the laser speckle velocimetry, which was developed originally in the area of solid mechanics. For further information on the historical development of PIV, the reader is referred to the review of Adrian [2]. Nowadays, different PIV configurations exist based on the measurement task. To investigate the two-dimensional (2D) velocity field in a plane, planar PIV is used, where the motion of particles is captured by one camera orthogonally oriented to that plane. If instead of one orthogonal camera, two cameras inclined by a predefined angle relative to the plane of interest are used, also the out of plane velocity component can be reconstructed. This setup is called stereoscopic PIV [125]. The extension to three-dimensional (3D) flows, where the particle motion is observed in a volume, is called tomographic PIV [104]. In TC flows, planar 2D PIV in the radial-axial plane was firstly applied by Wereley and Lueptow [139, 140]. Henceforth, this measurement technique has been used in many different configurations in TC flows, like planar 2D PIV in the radial-azimuthal plane [60, 128, 133], stereoscopic PIV in the radial-axial plane [99] or tomographic PIV [123, 124]. Within this thesis, planar PIV is performed in horizontal planes at different cylinder heights.

5.3.1 Fundamentals of planar 2D PIV

A planar 2D PIV setup consists of a light source that illuminates the tracer particles, a light sheet optic that expands the Gaussian laser beam to a planar plane, a camera oriented orthogonal to the plane that captures the scattered light of the particles, and the associated equipment for data storage and processing, according to Figure 5.6. Here, the focus is set to the so-called double-frame mode, where a pair of two subsequent particle images is taken at a predefined time separation Δt and triggered synchronously with a pulsed light source. PIV measurements strongly depend on the choice of the tracer particles. Their displacement over a specific time is taken as a measure for the fluid velocity. Therefore, the tracers have to follow the flow without slip, leading to the requirement of small ($d_p \approx 10\,\mu$m for water, see Westerweel *et al.* [141]) and neutrally buoyant particles ($\rho_p - \rho \approx 0$). The index p means particle and ρ is the fluid density. d_p represents the diameter of a spherical tracer. The following capability of small spherical particles to changes in the fluid velocity can be expressed in terms of the Stokes number St [125]. A particle is believed to follow the flow accurately when the particle response time τ_p is much smaller than the fluid time scale τ_f:

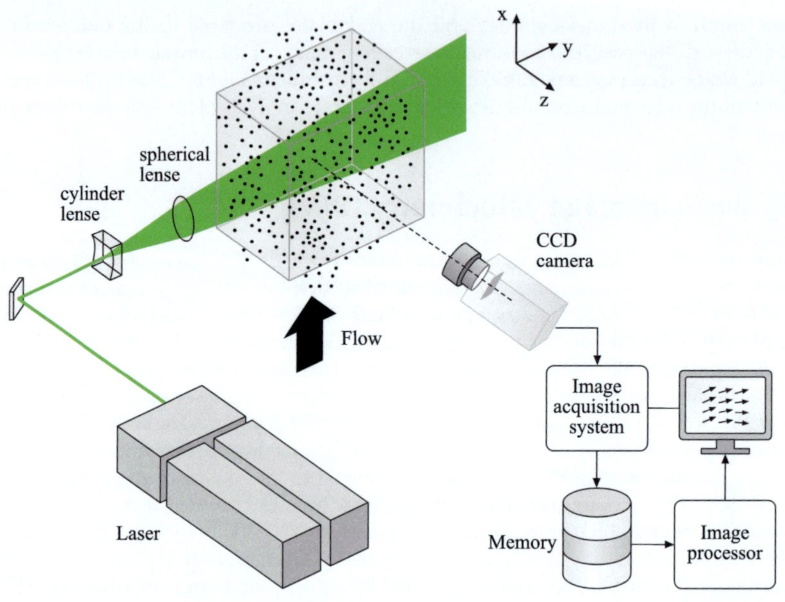

FIGURE 5.6: Schematic sketch of a classical planer 2D PIV measurement system, adopted from Tropea *et al.* [125].

$$St = \frac{\tau_p}{\tau_f} = \left(\frac{\rho_p}{\rho} - 1\right) \frac{d_p^2 u_S}{18\nu d} \ll 1. \tag{5.2}$$

In addition, the particles have to scatter the light of the light source sufficiently to obtain high-quality particle images. The light source is used to illuminate the area of interest twice within the separation time, while the duration of each illumination has to be very short to ensure sharp particle images. Further, the light intensity must be high to detect the scattered light and the generated light sheet thickness has to be small to assume a quasi 2D flow ($\mathcal{O}(0.1 - 2\,\text{mm})$, according to Nitsche and Brunn [80]). These requirements are fulfilled when using a solid-state frequency-doubled double-headed Nd-YAG laser with a wavelength of 532 nm. The pulsed, monochromatic and collimated light beam can be easily shaped to a thin light sheet by the combination of cylindrical and spherical lenses, the pulse separation is freely adjustable due to the two laser heads, and the pulse energy and pulse duration time are of order $\mathcal{O}(10\,\text{mJ} - 1\,\text{J})$ and $\mathcal{O}(5 - 15\,\text{ns})$, respectively [125]. In the last step of measurement, the scattered light of the particles is captured typically by a CCD camera (charged-coupled device), where the two subsequent particle images are stored on separate frames. The post-processing is used to calculate the velocity fields from the particle displacements Δx between the subsequent particle images. This analysis provides only reasonable results, if a sufficient quality of the particle images in terms of homogeneous seeding, uniform illumination and uniform image background is given. Thus, enhancement routines like filters etc. may be required to increase e.g. the signal-to-noise ratio. The calculation of the velocity field is based on a cross-correlation analysis. The images are divided into small sub-domains, the so-called interrogation areas (IA), and the cross-correlation is calculated for each IA of the first and second subsequent image. If the flow velocity is unequal to zero, the cross-correlation function exhibits a peak shifted from the origin, which represents the average particle displacement within the IA. Thus, the choice of the IA size defines the spatial resolution of the resulting velocity field. The velocity is given by

$$u = C\frac{\Delta x}{\Delta t}. \tag{5.3}$$

The constant C can be understood as the magnification of the optical setup and is determined in a calibration procedure. These days, there are a lot of optimization possibilities, to enhance the speed and accuracy of the mentioned PIV algorithm, of which only a few are mentioned in the following. For more information, the reader is referred to Tropea *et al.* [125]. To reduce computational time, the cross-correlation can be calculated in the Fourier space using the fast Fourier transform. When the particle displacement is determined at sub-pixel level, the resulting accuracy is better than 0.1 px. To capture large and small velocities in a turbulent flow, the particle displacement can be calculated several times, while the IA size is progressively decreased. This algorithm is called Multigrid-Analysis. To enable an optimal PIV analysis of a flow, Keane and Adriane [62, 63, 64] worked out four design rules for the experimental parameters based on Monte-Carlo simulations, which can be used as a guideline:

- Number of particle images per IA: 10-25

- Out-of-plane displacement should be smaller than 1/4 of the sheet thickness

- In-plane-displacement should be smaller than 1/4 of the IA length

- Displacement variations over IA should be smaller than 5% of its average

5.3.2 Basic setup

Within this thesis, two PIV measurement campaigns have been performed, one in the TvTCC and the other one in the BTTC experiment. While the basic setup was the same, the used components like lasers or CCD cameras as well as the software for the analysis were different. Therefore, here only the setup is shown and the specific components are described in the corresponding sections. The PIV setup is shown in Figure 5.7, similar to the one used by van der Veen *et al.* [128]. The laser light sheet is introduced into the gap through the transparent outer cylinder to illuminate the radial-azimuthal plane. The scattered light of the particles is captured by a CCD-camera that is focused on the light sheet through the transparent top plate. This arrangement avoids distortions of the particle images due to curved surfaces, as the top plate is plane. The laser is mounted on a traverse to measure the velocity field at different cylinder heights H and is shifted by constant steps, covering an axial length larger than two times the gap width d. The resulting field of view (FoV) contains only a small fraction of the gap concerning the azimuthal coordinate and the whole gap concerning the radial coordinate.

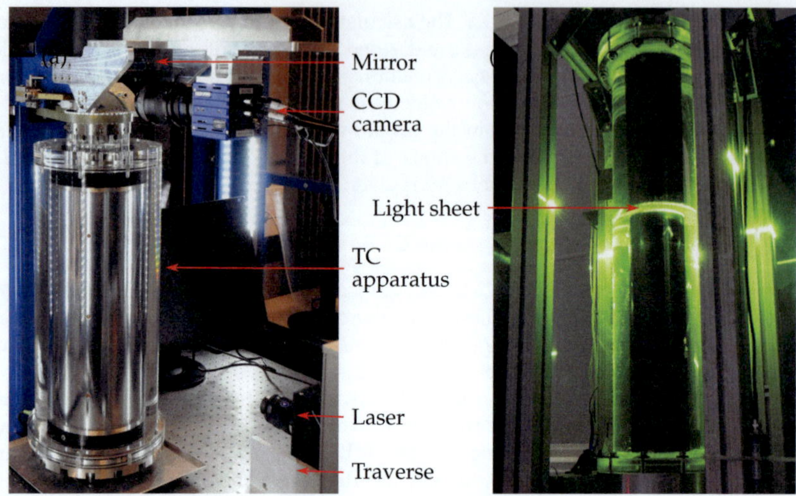

FIGURE 5.7: Photographs of the TC experiments with the PIV setup. (a) BTTC experiment including laser on a traverse, a CCD camera and a mirror. (b) TvTCC experiment with depicted horizontal laser light sheet.

When the cross-correlation is calculated, the particle displacement is given in pixels and Cartesian coordinates. The transformation to cylindrical coordinates can be done using equation (2.1), and the time separation of the two subsequent particle images Δt is predefined for the measurements. According to equation (5.3), only the constant C is missing to calculate the fluid velocity. The calibration process to quantify C is described in the next section. After calibration and coordinate transformation, the PIV measurements in the horizontal planes at different cylinder heights result in the radial u_r and azimuthal velocity component u_φ:

$$\mathbf{u} = u_r(r, \varphi, z, t)\mathbf{e}_r + u_\varphi(r, \varphi, z, t)\mathbf{e}_\varphi \tag{5.4}$$

It is important to note, that the velocity data for the different axial positions (z-coordinate) are measured sequentially, which only enables a statistical analysis in the z-coordinate direction. The azimuthal velocity component can be converted into the angular velocity $\omega = u_\varphi/r$ or the angular momentum $L = ru_\varphi$. To normalize these three quantities (denoted with a tilde symbol), they are mapped to an interval between 1 at the inner (r_1) and 0 at the outer cylinder (r_2) by use of the corresponding quantities at the cylinder walls:

$$\tilde{\omega} = (\omega - \omega_2)/(\omega_1 - \omega_2), \tag{5.5}$$
$$\tilde{u}_\varphi = (u_\varphi - u_{\varphi,2})/(u_{\varphi,1} - u_{\varphi,2}), \tag{5.6}$$
$$\tilde{L} = (L - L_2)/(L_1 - L_2). \tag{5.7}$$

Further, the radial velocity is normalized with the shear velocity u_S as:

$$\tilde{u}_r = u_r/u_S. \tag{5.8}$$

Similar, for other velocity-related quantities, like the standard deviation σ, the skewness S or kurtosis K of the velocity components as well as the kinetic energy E_{kin}, u_S is the appropriate parameter for normalization. In the style of w, u_φ and L, the radial coordinate r is mapped to an interval of 0 at the inner and 1 at the outer cylinder:

$$\tilde{r} = (r - r_1)/(r_2 - r_1). \tag{5.9}$$

These normalizations are used throughout the thesis.

5.3.3 Calibration

The calibration procedure is illustrated for the PIV campaign in the BTTC facility, where $r_1 = 75\,\text{mm}$, $r_2 = 105\,\text{mm}$ and $\eta = 0.714$. It has already been published in Froitzheim *et al.* [42]. A regular calibration using a calibration target was not applicable in this study, which is why the cross-correlation algorithm for the double images is performed in image coordinates $\mathbf{x}_{pic} = [x_{pic}, y_{pic}]^T$ (units in px), resulting in image velocities $\mathbf{u}_{pic}$ (units in px-displacement). Afterward, a transformation based on an in-situ image-based calibration is done, which results in the real coordinates $\mathbf{x}$ and velocities $\mathbf{u}$ (units in m and m/s). Note that due to the PIV algorithm, the procedure is performed in Cartesian coordinates and includes two steps. First, the positions of the IC ($\mathbf{x}_{pic,1}$) and OC ($\mathbf{x}_{pic,2}$) are determined in

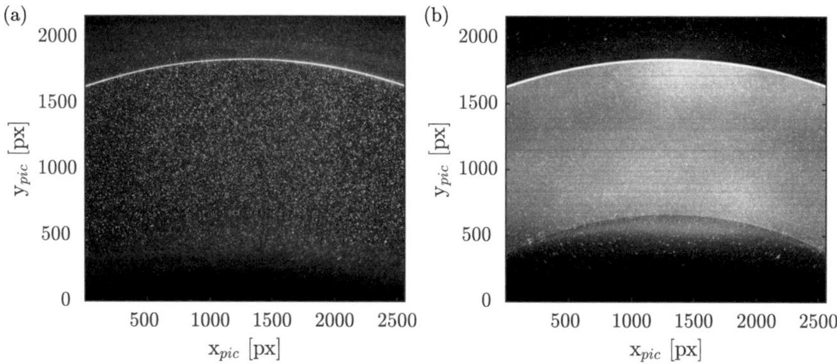

FIGURE 5.8: (a) Instantaneous snapshot of a raw particle image. (b) Standard deviation of 1500 particle images. Images are acquired in the radial-azimuthal plane during PIV measurements in the BTTC experiment.

image coordinates based on the local intensity change in the raw particle images. In the second step, concentric circles are fitted through these cylinder positions. In Figure 5.8 an instantaneous particle image and the standard deviation of the intensity values, calculated over 1500 snapshots, are depicted. While the outer cylinder is clearly visible as a bright, thin line, the inner cylinder is blurred (see Figure 5.8(b)). This difference in appearance results from the different cylinder materials (see Section 4.2) and the use of a bandpass filter in front of the CCD camera. Therefore, the position of the OC is given by the global intensity maximum, while the IC looks like a weak, local maximum. To determine the position of the IC, a circle is first fitted to the outer cylinder. Due to the known real radius ratio of $\eta = 0.714$, the inner cylinder position in the image can be estimated to be in between the circles with radii $\eta r_{pic,2} \pm s_{pic}$. If the constant is set to $s_{pic} = 20\,\text{px}$, the IC is given by the global intensity maximum in this short-banded area.

First of all the fitting algorithm of a single circle to the OC position $\mathbf{x}_{pic,2}$ with a total number of points $i = 1 : m$ is described, according to Gander *et al.* [46]. The vector $\mathbf{g}_{circ} = (c_1, c_2, r_{pic,2})^T$ is assumed to be the unknown center coordinates $\mathbf{c} = [c_1, c_2]^T$ and radius of the circle. Thus, a non-linear least-square problem has to be solved by minimizing the squared distance $d_{circ,i}$ of each point to the center:

$$\sum_{i=1}^{m} d_{circ,i}(\mathbf{g}_{circ})^2 = \sum_{i=1}^{m} (||\mathbf{c} - \mathbf{x}_i||_2 - r)^2 = \min. \tag{5.10}$$

The indices of $r_{pic,2}$ and $\mathbf{x}_{pic,2}$ are skipped for better visibility. An iterative solution can be calculated using a Taylor expansion around the solution vector $\hat{\mathbf{g}}_{circ}$ with a correction vector $\mathbf{h}_{circ}$, leading to

$$\mathbf{J}_{circ}\left(\hat{\mathbf{g}}_{circ}\right)\mathbf{h}_{circ} = -d\left(\hat{\mathbf{g}}_{circ}\right). \tag{5.11}$$

$\mathbf{J}_{circ}$ represents the Jacobian matrix defined by the partial derivatives $\partial d_{circ,i}\left(\mathbf{g}_{circ}\right)/\partial d_{circ,j}$, resulting in

$$\mathbf{J}_{circ} = \begin{bmatrix} \dfrac{c_1 - x_{11}}{||\mathbf{c} - \mathbf{x}_1||_2} & \dfrac{c_2 - x_{12}}{||\mathbf{c} - \mathbf{x}_1||_2} & -1 \\[2ex] \vdots & \vdots & \vdots \\[2ex] \dfrac{c_1 - x_{m1}}{||\mathbf{c} - \mathbf{x}_m||_2} & \dfrac{c_2 - x_{m2}}{||\mathbf{c} - \mathbf{x}_m||_2} & -1 \end{bmatrix} = \begin{bmatrix} \mathbf{J}_{circ}^{p} & -\mathbf{1} \end{bmatrix}, \tag{5.12}$$

and $\mathbf{J}_{circ}^{p}$ contains the first two columns of the Jacobian matrix. The following iteration persists of solving equation (5.11) based on a guessed start solution $\hat{\mathbf{g}}_{circ,0}$, and then updating the approximation $\hat{\mathbf{g}}_{circ,1} = \hat{\mathbf{g}}_{circ,0} + \mathbf{h}_{circ}$ according to the Gauss-Newton method, until the norm of $\mathbf{h}_{circ}$ undergoes a specific limit. This algorithm can be extended to a fitting procedure of two concentric circles with the solution vector $\mathbf{g}_{conc} = \left(c_1, c_2, r_{pic,1}, r_{pic,2}\right)^T$, when the Jakobian matrix $\mathbf{J}_{circ}$ and the distance function $\mathbf{d}_{circ}$ are adapted as

$$\mathbf{J}_{conc}\left(\mathbf{g}_{conc}\right) = \begin{bmatrix} \mathbf{J}_{circ,IC}^{p} & -\mathbf{1} & -\mathbf{0} \\[1ex] \mathbf{J}_{circ,OC}^{p} & -\mathbf{0} & -\mathbf{1} \end{bmatrix} \qquad \mathbf{d}_{conc}\left(\mathbf{g}_{conc}\right) = \begin{bmatrix} \mathbf{d}_{circ}\left(\mathbf{g}_{circ}\left(c_1, c_2, r_{pic,1}\right)\right) \\[1ex] \mathbf{d}_{circ}\left(\mathbf{g}_{circ}\left(c_1, c_2, r_{pic,2}\right)\right) \end{bmatrix}. \tag{5.13}$$

The results concerning the detection of the inner and outer cylinder as well as the concentric circle fit are depicted in Figure 5.9. The determined IC and OC positions are correctly located on the areas of high intensity, where the cylinders are displayed. The fitted concentric circles again reproduce these points. The ratio of fitted inner and outer cylinder radius is $r_{pic,1}/r_{pic,2} = 0.715$ very close to the real radius ratio of $\eta = 0.714$. If a similarity transformation is assumed between the image and the real world, the transformation from the image coordinates and velocities to the real ones is given by

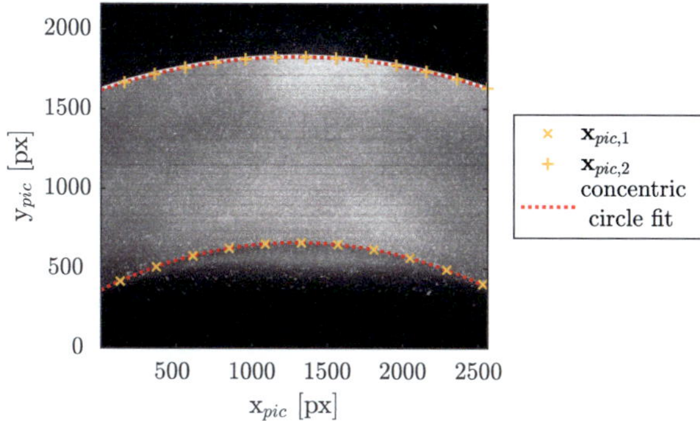

FIGURE 5.9: Example of PIV calibration based on a concentric circle fit. The standard deviation of 1500 particle images in the radial-azimuthal plane is shown with the determined locations of the IC and OC as well as the concentric circle fit.

$$\mathbf{x} = \begin{bmatrix} C & 0 \\ 0 & C \end{bmatrix} \left(\mathbf{x}_{pic}^{T} - \mathbf{c}^{T} \right), \tag{5.14}$$

$$\mathbf{u} = \begin{bmatrix} C & 0 \\ 0 & C \end{bmatrix} \frac{\mathbf{u}_{pic}^{T}}{\Delta t}, \tag{5.15}$$

$$C = \frac{r_2 - r_1}{r_{pic,2} - r_{pic,1}}. \tag{5.16}$$

C is the constant of equation (5.3) and can be called scale factor or magnification. In addition, with the determination of the IC and OC positions, the regions outside the gap can be easily masked out.

5.3.4 Measurement error

The quality of a measurement can be quantified by the measurement error. Within this section, three aspects of the error are discussed, namely, errors induced by misalignment of the camera relative to the light sheet, the deviation of the flow from the theoretical laminar solution for solid body rotation, and uncertainties due to sub-pixel interpolation for the detection of the correlation peak. Possible misalignment of the optical path becomes visible during the calibration process. As shown in the previews section for one specific axial height, the radius ratio of the facility and of the particle image are identical. However, measurements have been performed at different axial locations H, leading to different scale factors. In case of an ideal alignment, the scale factor should be a linear function of H. In Figure 5.10(a) the scale factor is plotted against the normalized height position for 23 different axial locations, separated by 4 mm. All points are almost on a line with a correlation coefficient of 0.9996. Thus, it can be concluded that the alignment of the camera and the light sheet and possible other distortions of the optical path are negligible. The small variations of the points from the line are positive as well as negative, indicating the absence of a systematic error.

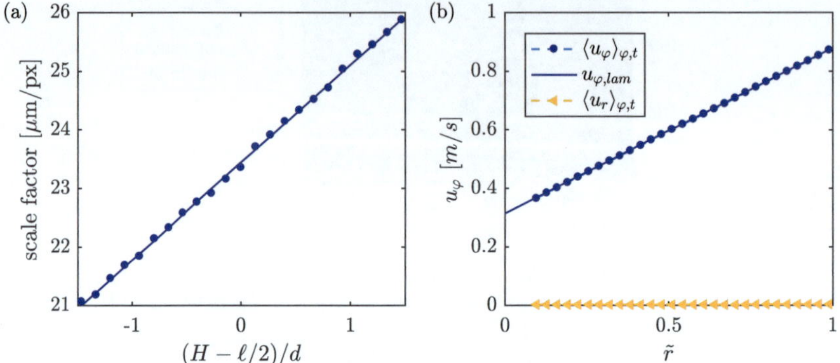

FIGURE 5.10: Example of the measurement accuracy of the PIV setup. (a) Scale factor for 23 different height positions (in BTTC). (b) Comparison of PIV results for solid body rotation at cylinder speeds $n_1 = n_2 = 120\,$rpm with the laminar Couette solution (in TvTCC).

In addition, a PIV measurement is performed in the TvTCC experiment for solid body rotation at mid-height with the PIV setup shown in Figure 5.7, to compare the result with the laminar Couette solution and quantify the absolute measurement error. According to Figure 5.10(b), the measured azimuthal velocity and the laminar Couette solution are in very good agreement with a maximum relative error of 0.8 %. The radial velocity component, which should vanish in the laminar case, has a maximum value of 3.6 mm/s. Compared to the azimuthal velocities of the inner and outer cylinder of $u_{\varphi,1} = 0.31\,$m/s and $u_{\varphi,2} = 0.88\,$m/s, this means a fraction of only 1.2 % and 0.4 %, respectively. Obviously, the real velocities are captured very accurately.

The error for the detection of the correlation peak, when a sub-pixel interpolation scheme is used, has a maximum value of $\Delta_{SM} = 0.1\,$px for a single measurement according to Nobach and Bodenschatz [81]. The standard error of the mean follows as $\Delta_{mean} = \Delta_{SM}/\sqrt{N}$ with N representing the number of spatial grid points or time steps used for the average. This quantity is analyzed when the PIV results are presented.

Chapter 6

Global angular momentum transport and pattern formation at $\eta = 0.357$

This chapter provides experimental and numerical results for the transport of angular momentum in a wide-gap configuration of $\eta = 0.357$. The experimental investigations are performed in the TvTCC facility, where the torque is measured as function of shear ($4.5 \times 10^3 \leq Re_S \leq 1.2 \times 10^5$) and rotation ($-0.5 \leq \mu \leq 0.2$), and the corresponding global flow pattern is revealed near the outer cylinder wall based on flow visualization. In addition, the experiments are compared with DNSs provided by Rodolfo Ostilla-Mónico to connect the properties of the torque to the velocity field. At the beginning, short overviews of the experimental setup and the numerical method are given in Sections 6.1 and 6.2, before the effective torque scaling with Re_S for $\mu = 0$ is discussed in Section 6.3. Thereafter, the momentum transport is analyzed for various μ and the results are compared to experimental data for a smaller gap of $\eta = 0.5$ from Merbold et al. [78] in Section 6.4. To illustrate the flow organization, that causes the specific dependency of the momentum transport on μ, the flow patterns are illustrated using the flow visualization technique in Section 6.5. At the end of this Chapter in Section 6.6, a conclusion is given, including a comparison of the results of this Chapter with other findings concerning the momentum transport in medium- and wide-gap TC flows. Note, that the results of this Chapter have been published in the journal Physical Review Fluids and the text is mainly taken from that paper: A. Froitzheim, S. Merbold, R. Ostilla-Mónico and C. Egbers, *Angular momentum transport and flow organization in Taylor-Couette flow at radius ratio of $\eta = 0.357$*, Phys. Rev. Fluids 4 (2019), 084605 [41].

6.1 Experimental setup and measurement procedure

The experiments have been carried out in the TvTCC experiment with an inner cylinder radius of $r_1 = 25$ mm, an outer cylinder radius of $r_2 = 70$ mm and a length of both cylinders of $\ell = 700$ mm. This setup results in a gap width of $d = 45$ mm, a radius ratio of $\eta = 0.357$ and an aspect ratio of $\Gamma = 15.6$. The torque is measured over the whole length of the inner cylinder with the end plates co-rotating with the outer cylinder. As working fluids, silicone oils of different viscosity, namely M3, M5, M10 and M20, and air are used. The torque values for an air-filled gap are subtracted from the ones using the silicone oils, to correct the torque signal for bearing friction. In addition, the temperature of the fluid is monitored to get access to the actual fluid density and viscosity. More detailed information about the experimental setup and the torque measurement technique can be found in the Sections 4.1 and 5.2, respectively. For the torque measurements within this chapter, a fixed protocol was used. It can be distinguished between measurements for $\mu = 0$, where the IC speed is increased sequentially, and measurements at a fixed shear Reynolds number Re_S, where the ratio of angular velocities μ is changed sequentially. In both cases, the cylinder speed of the first measurement point is set with a subsequent waiting time of approximately 10 min to let the flow evolve. Afterward, the torque is measured for 90 s at 10 Hz, before the cylinder speeds are adapted for the next measurement point. Here, the change in cylinder speed is small and the required waiting time can be reduced to 120 s. Due to this waiting time, the measurements can be assumed to be quasi-stationary. In the case of pure inner cylinder rotation, the fluid temperature is measured for each data point directly before the measurement phase. In case of differential rotating flow states, the inner cylinder speed is adapted always before the outer one and for every fourth data point, the outer cylinder speed is set to zero and the fluid temperature is measured. For the purpose of flow visualization, the flow is seeded with aluminum flake particles and blue pigments, and is illuminated by two halogen lamps. The motion of these tracers is recorded by an *Optronis* CR 3000x2 high-speed camera with 1690×1710 px and only a central vertical line of the acquired intensity information is analyzed to reveal the flow structure according to Section 5.1. For each visualized flow state, a waiting time of 10 min was used before the images of the flow were taken. Each visualization consists of 2900 images, recorded at a frequency of 60 Hz.

6.2   Numerical Simulations

To complement the experiments, DNSs of TC flow are performed. Therefore, the incompressible Navier-Stokes equation is solved in cylindrical coordinates in a rotating reference frame according to equation (2.18). Spatial discretization is achieved by the use of a second-order energy-conserving centered finite difference scheme, and time-marching is performed with a low-storage third-order Runge-Kutta for the explicit terms and a second-order Adams-Bashworth scheme for the implicit treatment of the wall-normal viscous terms. Further details of the algorithm can be found in [126, 134]. This code has been previously used and validated extensively for Taylor-Couette flow [86].

Unlike the experiments, the numerical simulations use axially periodic boundary conditions and the aspect ratio is fixed to $\Gamma = 2$. For this aspect ratio, exactly one vortex pair with wavelength $2\lambda_{TV} = 2d$ fits into the simulated volume. Due to the periodic boundary conditions, this corresponds to the solution for infinitely long cylinders as long as periodicity effects are negligible. Indeed, it is known that one single roll is sufficient to accurately capture most statistics, including torque [14, 84]. The non-dimensional radius ratio η is also matched to that of the experiment to $\eta = 0.357$. Unlike previous studies, no additional rotational symmetry is imposed, and the full 2π azimuthal extent of the domain is simulated. Due to the wide gap, the streamwise domain length is already quite small at the inner cylinder as compared to the gap size and becomes at r_1 approximately $3.5d$. According to Ostilla-Mónico *et al.* [85], a minimum streamwise domain length of πd is needed to produce accurate decorrelations, so it can be suspected that introducing a degree of rotational symmetry, which reduces this domain length, will introduce artifacts. The simulations are performed in the reference frame of Dubrulle *et al.* [31] such that both cylinders rotate with opposite velocities. In this frame, the two control parameters naturally become the shear Reynolds number and the rotation number defined in Section 2.2.3.

The shear Reynolds number is varied between $Re_s = 5 \times 10^3$ and $Re_s = 4 \times 10^4$, with resolutions ranging from $N_\varphi \times N_r \times N_z = 256 \times 256 \times 512$ to $768 \times 384 \times 768$, in the azimuthal, radial and axial directions, respectively. The axial and azimuthal grid distributions are homogeneous, but points are clustered in the radial direction. Due to the asymmetry of the cylinders, higher resolutions are needed at the inner cylinder to properly resolve the structures. For the radial direction, this is solved by clustering more points. The azimuthal discretization becomes larger with increasing radius, which naturally provides for higher resolution at the inner cylinder. The axial direction can become problematic, as it is discretized in a homogeneous manner. In this direction, the resolution must be set by the structures at the inner cylinder, and thus the outer cylinder wall is over-resolved in this direction. At low Reynolds numbers, the resolution is chosen such that dispersive effects cannot be observed when looking at the flow field. For higher Reynolds numbers, the resolution is measured in viscous units, i.e. normalized by $\delta_v = \nu/u_\tau$, where $u_\tau = \sqrt{\tau_W/\rho}$ is the frictional velocity, with the shear at the wall $\tau_W = \mathcal{T}/(2\pi r^2 \ell)$. The viscous unit at the inner wall is chosen as it is more restrictive due to the higher shear τ_W. The spatial discretization is then approximately $r\Delta\varphi^+ \approx 10$, $\Delta z^+ \approx 4$ and $\Delta r^+ \in (0.5, 5)$ in inner wall units, as in [88]. The normalization procedure exemplified for the radial coordinate at the inner cylinder is $r^+ = (r - r_1)/\delta_v$.

The rotation parameter R_Ω is varied between $R_\Omega \in (0.08, 0.84)$, corresponding to values of $\mu \in (-0.3, 0.075)$. Temporal convergence is assessed by measuring the difference in torque between both cylinders, and ensuring that the time-average of both torques coincides within 3%.

6.3   Scaling of torque with Re_S for $\mu = 0$

At first, the effective scaling of the torque with the Reynolds number is studied, where usually a power-law ansatz of the form $\mathrm{Nu}_\omega \sim Re_S^{\alpha-1}$ is assumed. The torque for pure inner cylinder rotation is measured over a range of $4.5 \times 10^3 \leq Re_S \leq 1.2 \times 10^5$ using several working fluids. In Figure 6.1(a) the non-dimensionalized torque, i.e. the Nusselt number as a function of the shear Reynolds number is shown for all used working fluids in logarithmic scale. The experimental data points across working fluids only scatter slightly in a small area without observable discontinuities at the crossovers. In addition, the numerical data points are in good agreement with the experimental results. As could be expected, the non-dimensional momentum transport increases with an increasing shear Reynolds number. In a first step, the Nusselt number is compensated by $Re_S^{-0.65}$. This scaling exponent was found for $\eta = 0.5$ by Merbold *et al.* [78] in the same Reynolds number range as this study (see Figure 6.1(b)). The compensated Nusselt number is approximately constant up to $Re_S \leq 1.3 \times 10^4$ and decreases for higher shear Reynolds numbers. This behavior suggests that some kind of transition takes place, changing the overall momentum scaling. To precisely characterize the transitions, the local exponent α of the power-law ansatz $\mathrm{Nu}_\omega \sim Re_S^{\alpha-1}$ is calculated. As described by Lathrop *et al.* [66], α is defined as:

$$\alpha(Re_S) = \frac{\partial \left(\log_{10} \mathrm{Nu}_\omega \right)}{\partial \left(\log_{10} Re_S \right)} + 1. \tag{6.1}$$

Instead of directly calculating the derivative of the Nu_ω-Re_S-curve, which may lead to difficulties due to the mentioned slight scattering of the data points and the existence of more than one data point at a specific Reynolds number, a linear least square fit is computed across equidistant intervals $\Delta_{10}(Re_S)$ in logarithmic scale, centered by the individual points. The scaling exponent, obtained by taking an interval of $\Delta_{10}(Re_S) = 0.5$, is shown in Figure 6.1(c).

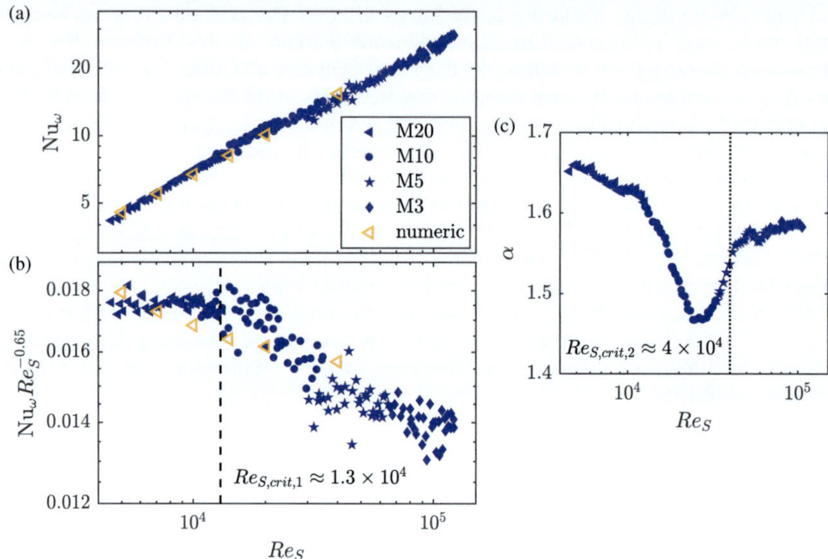

FIGURE 6.1: (a) Nusselt number Nu_ω as a function of the shear Reynolds number Re_S for pure inner cylinder rotation $\mu = 0$. Different filled blue symbols represent experimental data for different working fluids. Yellow open symbols denote numerical results. (b) Compensated Nusselt number $\mathrm{Nu}_\omega Re_S^{-0.65}$ as a function of Re_S for $\mu = 0$. A transition appears at approximately $Re_{S,crit,1} \approx 1.3 \times 10^4$, which is marked as dashed line. (c) Corresponding local scaling exponent $\alpha - 1$ as a function of Re_S, calculated for a bin size of $\Delta_{10}(Re_S) = 0.5$. A transitional behavior is visible in the region $1.3 \times 10^4 \leq Re_S \leq 4 \times 10^4$. The dotted line indicates the end of this transition at $Re_{S,crit,2} \approx 4 \times 10^4$, where α starts to monotonically increase.

The exponent α has a slow downwards trend starting at $\alpha \approx 1.65$ up to $Re_S \approx 1.3 \times 10^4$, than strongly decreases to values around $\alpha \approx 1.47$, before it subsequently increases again noticeably. For shear Reynolds numbers above $Re_S > 4 \times 10^4$, the exponent is monotonically increasing and depicts a slight slope. Apparently, the effective momentum scaling reveals a transition confined by two critical Reynolds numbers, which are $Re_{S,crit,1} \approx 1.3 \times 10^4$ and $Re_{S,crit,2} \approx 4 \times 10^4$. As the value of α is always smaller than $5/3$, the here observed transition cannot be connected to the transition from the classical to the ultimate regime, which coincidentally happens at a very similar value for the shear Reynolds number in the case of $\eta \geq 0.714$ ($Re_{S,crit} = 1.04 \times 10^4$ or $Ta_{crit} = 3 \times 10^8$) [86]. It is further worth to mention, that the values of α calculated within this study are much

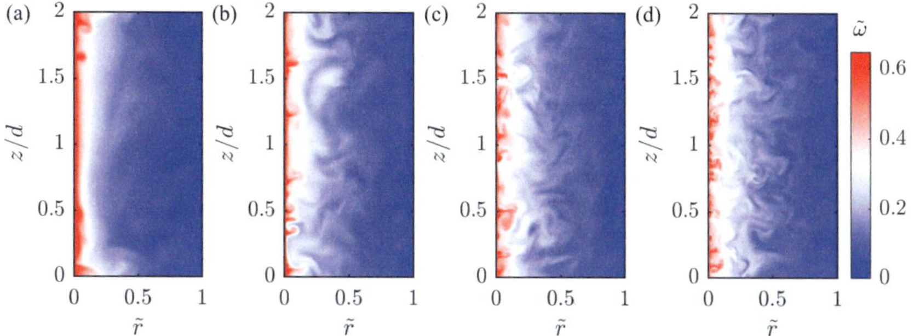

FIGURE 6.2: Constant-azimuth cut of normalized instantaneous angular velocity $\tilde{\omega}$ for four different Re$_S$ at $\mu = 0$. (a) Re$_S = 5 \times 10^3$, (b) Re$_S = 1 \times 10^4$, (c) Re$_S = 2 \times 10^4$ and (d) Re$_S = 4 \times 10^4$.

smaller than the one found by Burin *et al.* [19] for nearly the same η. They reported scaling exponents of $\alpha > 5/3$ for $Re_1 \geq 2 \times 10^4$ on the basis of local LDV measurements. This discrepancy, and in particular the question of whether the measurements of this study reach the ultimate regime or not, will be discussed in the further course.

With the numerical simulations the flow fields can be explored to uncover aspects of the transitions. In Figure 6.2 constant-azimuth cuts of the instantaneous angular velocity for increasing shear Reynolds numbers are shown. It can be seen that the structures at the inner cylinder appear to be smaller than the ones at the outer cylinder, as can be expected by the higher frictional Reynolds number ($Re_{\tau,2} = u_{\tau,2}d/(2\nu) = \eta Re_{\tau,1}$). Aside from the quiescent zone in the region of the inner cylinder for $Re_S = 5 \times 10^3$, which disappears at higher Re_S, no fundamental changes are occurring. The structures (plumes) of angular velocity become smaller and more numerous with increasing Re_S as could be expected. Note that in a similar manner, no large changes in flow topology were observed for $\eta = 0.5$ during the transitions by [86].

η	Re$_S$	μ	$Re_{\tau,1}$	$Re_{\tau,2}$
0.357	5×10^3	0	120	43
0.357	1×10^4	0	209	75
0.357	2×10^4	0	363	130
0.357	4×10^4	0	635	230
0.5	7.9×10^4	0	1112	556
0.714	1.4×10^4	0	252	180
0.909	1.4×10^4	0	240	220

TABLE 6.1: Overview of inner and outer cylinder frictional Reynolds number $Re_{\tau,1/2}$ for pure inner cylinder rotation in the transitional regime of this study and comparison with more narrow-gap investigations in the region of the classical-ultimate turbulent transition at $\eta = 0.5, 0.909$ from [86] and $\eta = 0.714$ from [83].

Focusing on the boundary layer, in table 6.1, the frictional Reynolds numbers of the present numerical simulations for $\mu = 0$ at the inner and outer cylinder are summarized. For comparison purposes, also the values of Re_τ for the appearance of the ultimate regime happens for $\eta = 0.5$, $\eta = 0.714$ and $\eta = 0.909$ are included (taken from Refs. [83, 86]). This appearance is triggered by a boundary layer transition, and thus a similar frictional Reynolds number could be expected for the transition. From the table it can be seen that there is a markedly increased value of Re_τ for the transition at $\eta = 0.5$. This was previously attributed to the effect of concavity in stabilizing the outer cylinder boundary layer [83]. Accordingly, the here observed transition for $\eta = 0.357$ in the range of $1.3 \times 10^4 \leq Re_S \leq 4 \times 10^4$ is unlikely caused by the development of turbulent BLs, especially at the outer cylinder, due to the too low shear and curvature stabilization. Something to note, however, is that the inner frictional Reynolds number $Re_{\tau,1}$ at $Re_{S,crit,1} \approx 1.3 \times 10^4$ has a similar value as $Re_{\tau,2}$ at $Re_{S,crit,2} \approx 4 \times 10^4$, which can be interpreted as follows. The observed transition takes place at a frictional Reynolds number of about $Re_{\tau,crit} \approx 230$. As this value is reached earlier at the inner than at the outer cylinder, some transition happens in two steps. These results show the pronounced asymmetry inside such a wide-gap TC flow, which is probably the reason for the strong discrepancy between the here presented globally measured effective scaling exponent and the locally measured one by Burin *et al.* [19].

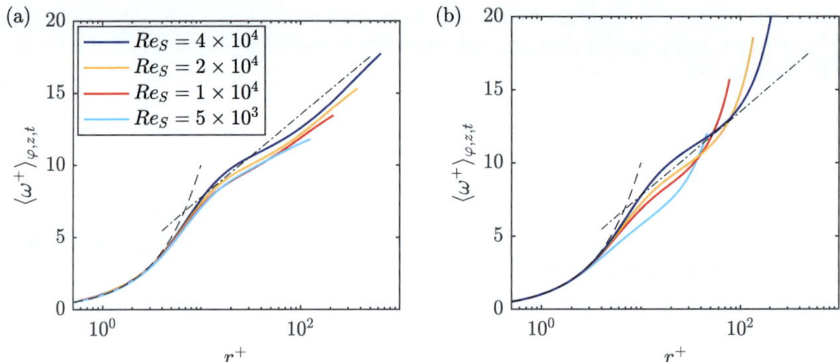

FIGURE 6.3: Azimuthally, axially and temporally averaged normalized angular velocity profiles of (a) $\omega_1^+(r_1^+)$ near the inner cylinder for radii $\tilde{r} \in [0, 0.5]$ and (b) $\omega_2^+(r_2^+)$ near the outer cylinder for radii $\tilde{r} \in [0.5, 1]$. The angular velocities are defined as $\omega_1^+ = (\omega(r_1) - \omega(r))/(u_{\tau,1}/r_1)$ and $\omega_2^+ = (\omega(r) - \omega(r_2))/(u_{\tau,2}/r_2)$, while the radial coordinates are calculated by $r_1^+ = (r - r_1)/\delta_{v,1}$ and $r_2^+ = (r_2 - r)/\delta_{v,2}$. The subscripts $1, 2$ are omitted in the labels. As guides for the eye, the dashed line represents the viscous sublayer $\omega^+ = r^+$ and the dash-dotted line a logarithmic law of the form $\omega^+ = 2.5 \ln(r^+) + 2$.

If the observed transition is not the transition to the ultimate regime, because it is inconsistent with the scaling law, then what is it? To further investigate the nature of the boundary layers that have been formed, the profiles of angular velocity are shown in Figure 6.3 for the inner and outer boundary layer in wall units. While the inner cylinder boundary layer could be showing some logarithmic behaviour with deviations due to curvature effects, the outer cylinder boundary is definitely of laminar type. The asymmetry between boundary layers is further explored in Figure 6.4, where the average and root-mean-squared profiles of angular velocity and angular momentum are shown. The differences of the angular velocity profiles with increasing shear Reynolds number are not very visible, so the focus of the analysis here is on the angular momentum. There, it can be seen that only for the highest Re_S achieved in the simulations, the mean angular momentum becomes equal to the arithmetic mean of 0.5. This provides an indication that the transition seen at $Re_{S,crit,2}$ is probably related to the outer cylinder's capacity to emit plumes at a rate that achieves marginal stability and equalizes the angular momentum in the bulk. It has to be stressed that not only is the shear smaller at the outer cylinder due to the fact that the radius is larger (see table 6.1), and thus the velocity gradient must be smaller to achieve the same torque, but there is also a concave curvature of the outer cylinder that stabilizes the emission of turbulent plumes. Thus, the shear required for plume emission becomes higher.

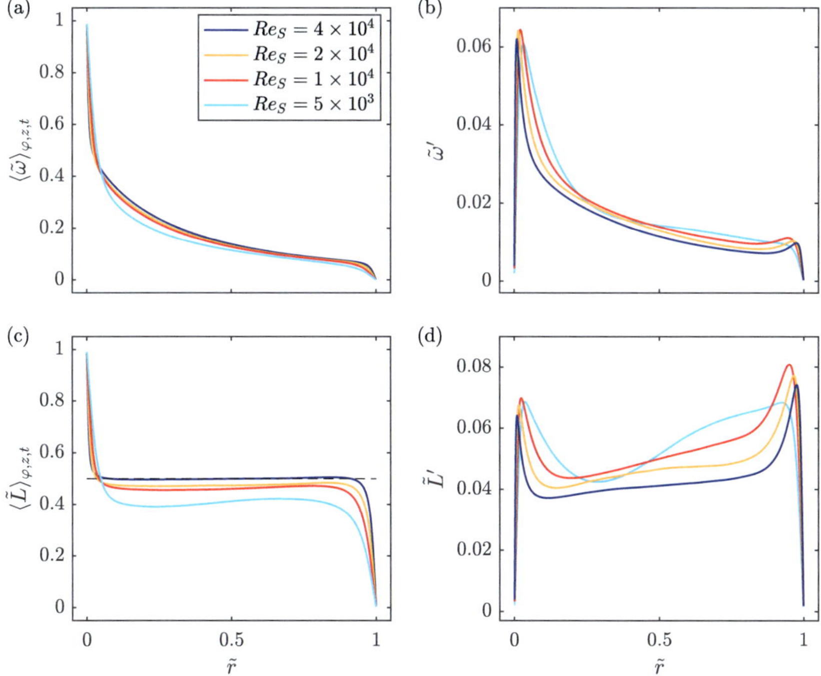

FIGURE 6.4: Azimuthally, axially and temporally averaged normalized (a) angular velocity $\langle \tilde{\omega} \rangle_{\varphi,z,t}$ and (c) angular momentum $\langle \tilde{L} \rangle_{\varphi,z,t}$ for $\mu = 0$ and varying Re_S. The corresponding root-mean-squared fluctuations of the normalized (b) angular velocity $\tilde{\omega}'$ and (d) angular momentum $\tilde{L}'$ are depicted on the right-hand side. Data are based on numerical simulations.

Further evidence of this is seen in Figure 6.4(b), where the root-mean-squared fluctuations of the angular momentum are shown. For $Re_S = 5 \times 10^3$ only a flat plateau at the outer cylinder is existing, indicating that there is no considerable production of fluctuations by the outer cylinder. Once Re_S increases, a sharp peak of fluctuations at the outer cylinder boundary layer appears, probably coinciding with the transition seen at $Re_{S,crit,1}$, even if the number is slightly different for the simulations. From the previous discussion, it appears that even once the peak arises, the turbulence level is not enough to equalize the angular momentum until $Re_{S,crit,2}$. Thus, it can be suggested that both transitions are essentially associated with the capacity of the outer cylinder to emit angular momentum plumes to develop a marginally stable profile.

6.4   Scaling of torque with μ for $Re_S = const$

Now, the dependence of the Nusselt number on the rotation ratio μ is analyzed when the shear Reynolds number is kept constant. In Figure 6.5, the combined numerical and experimental results are shown in the range of $5 \times 10^3 < Re_s < 2.5 \times 10^4$.

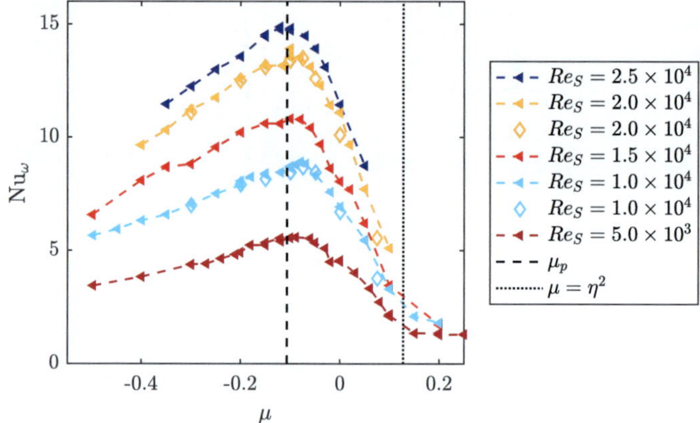

FIGURE 6.5: Nusselt number Nu_ω as a function of the rotation ratio μ for different shear Reynolds numbers Re_S. Filled symbols represent experimental data (already discussed in [77]) and open symbols numerical data. The dashed line indicates the prediction of the torque maximum according to Brauckmann and Eckhardt [15], while the dotted line indicates the Rayleigh stability criterion.

The Nusselt number exhibits for all Reynolds numbers a maximum in the low counter-rotating regime, which asymptotically comes closer to the prediction of Brauckmann and Eckhardt [15] with increasing Re_S (see equation (3.3)). The maximum location according to the angle bisector hypothesis ($\mu \approx -0.25$) lies noticeable far from the peak within the stronger counter-rotating regime. For $Re_S = 10^4$ and $Re_S = 2 \times 10^4$, where also simulations have been performed, the experimental and numerical data agree very well. The increase of the Nusselt number, coming from co-rotating cylinders to μ_{max}, is much steeper, than the decline for higher counter-rotation. This behavior results probably from the small distance of μ_{max} from the Rayleigh stability line at $\mu = \eta^2 = 0.128$ and will be revisited later. In the Rayleigh stability region, the flow should be laminar and the

Nusselt number has to be 1. Here, a nearly constant value of $\mathrm{Nu}_\omega \approx 1.3$ is found in the Rayleigh stable regime for the lowest investigated Reynolds number of $Re_S = 5 \times 10^3$. This value, which is slightly higher than expected, may be the result of end wall effects causing a large-scale circulation, the so-called Ekman vortices. Such a secondary flow can grow, especially in laminar flows, and fill the whole gap [7].

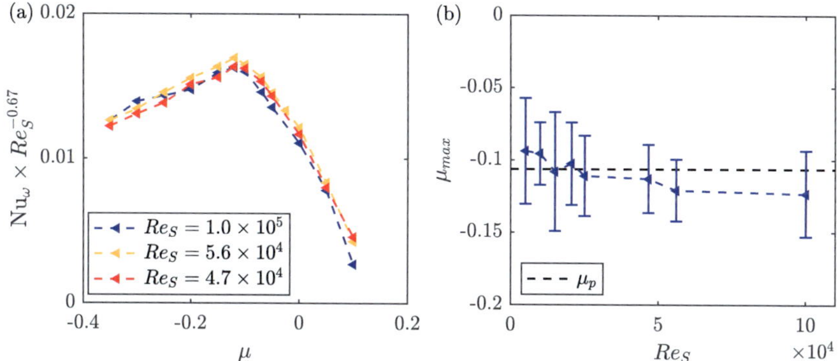

FIGURE 6.6: (a) Compensated Nusselt number $\mathrm{Nu}_\omega \times Re_S^{-0.67}$ as a function of the rotation ratio μ for three shear Reynolds numbers Re_S. (b) Evolution of the torque maximum location as a function of Re_S, calculated by a quadratic fit to the measured profiles in the range of $-0.16 \leq \mu \leq -0.06$. The dashed line represents the prediction μ_p of the torque maximum [15].

At higher shear Reynolds numbers, investigated only experimentally, the overall shape of the Nu_ω curve stays the same with a torque maximum close to μ_p, see Figure 6.6(a). To determine the exact location of the torque maximum, a quadratic curve of the form $\mathrm{Nu}_\omega = p_1\mu^2 + p_2\mu + p_3$ is fitted in the range of $-0.16 \leq \mu \leq -0.06$ to the measured profiles. This way the location of the torque maximum is given by $\mu_{\max} = -p_2/(2p_1)$ and its uncertainty is defined as [15]

$$\Delta \mu_{\max} = \sqrt{\frac{\Delta \mathcal{T}}{p_1 \mathcal{T}} (\mathrm{Nu}_\omega)_{\max}} \tag{6.2}$$

The location of $\mu_{\max}$ shifts with an increasing shear Reynolds number to more negative values and appears to settle down to a nearly fixed value for $Re_S \geq 5.6 \times 10^4$, reaching $\mu_{\max} = -0.123 \pm 0.030$ at $Re_S = 10^5$ (see Figure 6.6(b)). The prediction of Brauckmann and Eckhardt [15] is within the uncertainty range of the maximum location found within this study.

To finalize this section, the compensated torque in terms of R_Ω vs $\mathrm{Nu}_\omega \times Re_S^{-0.64}$ is shown. Dubrulle *et al.* [31] had already noted that the non-dimensional torque depends very little on the radius ratio η, and Brauckmann *et al.* [17] further developed this by showing a wide collapse of the compensated $\mathrm{Nu}_\omega(R_\Omega)$-curves for $R_\Omega > 0.25$ for radius ratios higher than $\eta = 0.5$, and for $R_\Omega > 0.1$ for radius ratios higher than 0.8. Brauckmann *et al.* [17] noted that the region where the large-gap Nusselt number collapses is where it is not affected by the radial flow partitioning. The appearance of radial flow partitioning introduces a strong dependence of the Nusselt number on the radius ratio. Further proof

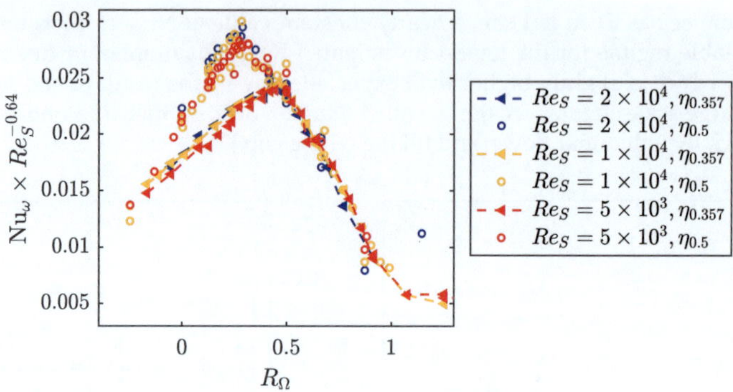

FIGURE 6.7: Comparison of the compensated Nusselt number $Nu_\omega Re_S^{-0.64}$ as a function of the rotation number R_Ω for the wide gaps $\eta = 0.357$ (experiments from this study) and $\eta = 0.5$ (taken from Merbold *et al.* [78]) for three shear Reynolds numbers Re_S. The curves collapse between the Rayleigh stability line ($R_\Omega = 1$) up to the appearance of radial flow partitioning.

of this for the wide-gap considered here can be seen in Figure 6.7. The curves for the compensated torque at three different Re_S and two radius ratios $\eta = 0.357$ (experiments from this study) and $\eta = 0.5$ (taken from Merbold *et al.* [78]) collapse in the interval between $R_\Omega(\mu_{max}(\eta = 0.357))$ and $R_\Omega = 1$. This corresponds to TC flow between slight counter-rotation, and co-rotation up to the Rayleigh stability line. For smaller values of R_Ω, radial flow partitioning arises and the η dependence of Nu_ω is no longer absent. The statements by Dubrulle *et al.* [31] and Brauckmann *et al.* [17] hold true even for large values of R_C.

6.5 Flow topology

The excellent agreement between the results of this thesis and the prediction of Brauckmann and Eckhardt [15] for the location of the maximum transport of angular momentum suggests, that the torque maximum is caused by strengthened large-scale Taylor rolls. To verify this explanation, flow visualizations have been performed as described in Section 5.1. The intensity distribution $\tilde{\mathcal{I}} = \mathcal{I} - \mathcal{I}_{lam}$ of an axial central line for the observed flow as a function of time t at the shear Reynolds number $Re_S = 2.5 \times 10^4$ and different μ is depicted in Figure 6.8. For co-rotation in the Rayleigh stable regime at $\mu = 0.2$, the flow is laminar, indicated by a homogeneous light intensity over time. However, slightly bright and inclined lines are visible especially below mid-height, which are crossing and may be fingerprints of end wall induced Ekman vortices. This would explain the measured Nusselt number of $Nu_\omega \approx 1.3$ in that regime. When the rotation ratio is decreased into the linearly unstable regime at $\mu = 0.1$, the flow becomes fully turbulent, resulting in strong fluctuations of the light intensity in the axial as well as in the temporal coordinate direction. The flow appears as a black background with bright spots, which are randomly distributed in the whole field of view. In case of pure inner cylinder rotation ($\mu = 0$), the intensity distribution appears similar to the previous one with small differences for the bright spots. Here, these spots are smeared into the direction of the temporal coordinate direction. When a rotation rate of $\mu = -0.1$ close to μ_{max} is adjusted, pronounced large-scale vortices are formed inside the flow, conforming with the theory of Brauckmann and Eckhardt [15]. Across the axial length of approximately six gap widths, three vortex pairs can be identified, which are stationary in time concerning their axial position. At even higher counter-rotation rates, the intensity distributions become again brighter, more homogeneous and the large-scale rolls disappear. This evolution is due to the stabilization of the flow in the outer gap region induced by a stronger rotation of the outer cylinder. As these videos only visualize a small flow region close to the outer cylinder wall, a probably further existence of the turbulent Taylor vortices in the inner gap cannot be shown.

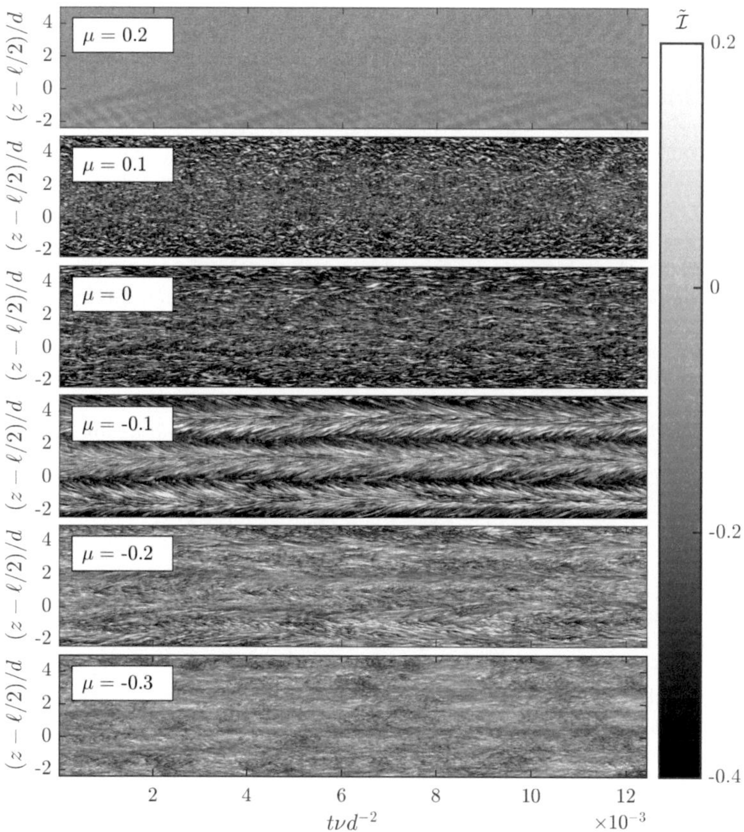

FIGURE 6.8: Space-time diagrams of the intensity distribution $\tilde{\mathcal{I}}$ along an axial, central line over time t for the shear Reynolds number $Re_S = 2.5 \times 10^4$ and different rotation ratios μ. All videos have been acquired at 60 Hz.

Also two additional flow states have been investigated in the region of the torque maximum at $\mu = -0.1$ for higher shear Reynolds numbers, depicted in Figure 6.9. Again,

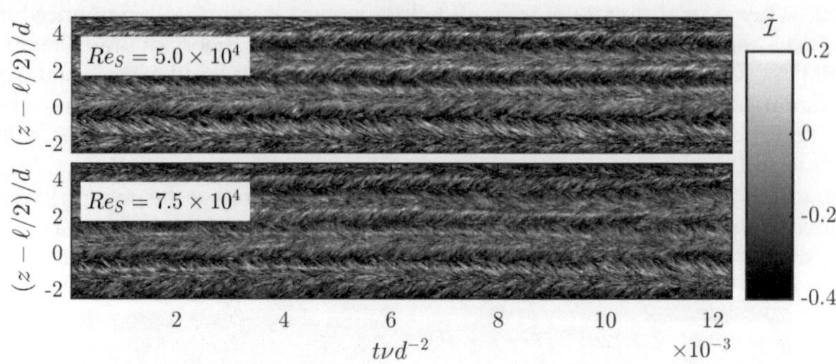

FIGURE 6.9: Space-time diagrams of the intensity distribution $\tilde{\mathcal{I}}$ along an axial, central line over time t for the shear Reynolds numbers of $Re_S = 5 \times 10^4$ and $Re_S = 7.5 \times 10^4$ for $\mu = -0.1$. Both videos have been acquired at 60 Hz.

pronounced large-scale rolls are found in the flow, which are stable in time. In contrast to the flow at $Re_S = 2.5 \times 10^4$, both cases exhibit four Taylor roll pairs over an axial length of approximately six gap widths. It should be noted that for the present investigations no fixed acceleration rate for the cylinders was used, whose variation can cause different flow states [74]. Also, a swap from three to four roll pairs or back could not be observed at a constant shear Reynolds number. Nevertheless, the present finding suggests that multiple flow states defined by the overall number of Taylor vortices filling the gap are possible also in this wide-gap TC geometry. To quantitatively specify the importance of turbulence and large-scale rolls in the flow, the two following quantities are calculated:

$$\mathcal{I}_{turb} = \left\langle \left(\tilde{\mathcal{I}} - \langle \tilde{\mathcal{I}} \rangle_t \right)^2 \right\rangle_{t,z},$$
$$\mathcal{I}_{LSC} = \sigma_z \left(\langle \tilde{\mathcal{I}} \rangle_t^2 \right). \tag{6.3}$$

$\mathcal{I}_{turb}$ is a measure for the fluctuations of the light intensity and therefore for the turbulence inside the flow. Besides, $\mathcal{I}_{LSC}$ quantifies the amplitude of the temporally averaged axial intensity variation, and therefore increases when large-scale rolls are formed in the gap. Both quantities are depicted in Figure 6.10(a) as a function of the rotation ratio μ. The intensity fluctuation $\mathcal{I}_{turb}$ strongly increases, when the rotation ratio is reduced starting at $\mu = 0.2$ in the Rayleigh stable regime into the unstable regime at $\mu = 0.1$. For a further decrease of μ, $\mathcal{I}_{turb}$ continuously decreases up to $\mu = -0.4$. On the other hand, $\mathcal{I}_{LSC}$ increases from $\mu = 0.2$ to $\mu = -0.1$. For smaller μ, a rapid breakdown of the amplitude is seen. Accordingly, the results of visual inspection of the space-time diagrams depicted in Figure 5.1 are supported quantitatively by the parameters $\mathcal{I}_{turb}$ and $\mathcal{I}_{LSC}$. Moreover, in Figure 6.10(b), the axial auto-correlation function $R_{\mathcal{I}\mathcal{I}}$ of the temporally averaged axial intensity profile is depicted for the three flow cases at $\mu = -0.1$ in the region of the torque maximum. The correlation coefficients depict a pronounced oscillation in the axial coordinate direction due to the large-scale Taylor rolls and the first minimum is a measure for the axial wavelength λ_{TV}. $\lambda_{TV}(Re_S = 2.5 \times 10^4) = 1.02d$, $\lambda_{TV}(Re_S = 5.0 \times 10^4) = 0.73d$

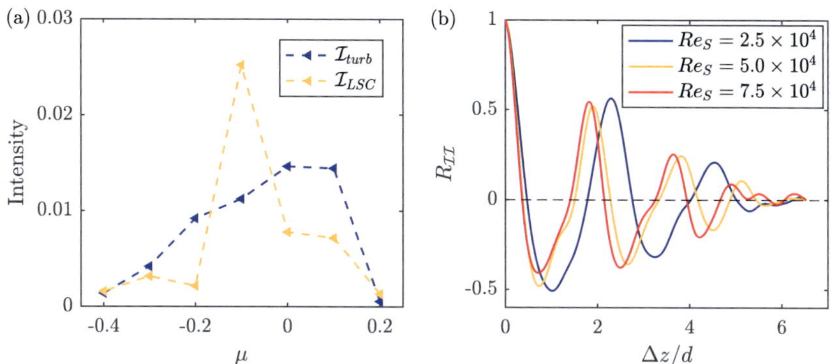

FIGURE 6.10: (a) Fluctuation of the light intensity $\mathcal{I}_{turb}$ and amplitude of the temporally averaged axial intensity variation $\mathcal{I}_{LSC}$ for $Re_S = 2.5 \times 10^4$ as a function of μ. (b) Axial auto-correlation coefficient R_{II} of the temporally averaged axial intensity profiles $\langle \tilde{\mathcal{I}} \rangle_t$ for $\mu = -0.1$.

and $\lambda_{TV}(Re_S = 7.5 \times 10^4) = 0.74d$ are found. For the lowest shear Reynolds number, the Taylor vortices capture nearly one gap width in the axial direction but become axially compressed at higher Re_S. The effect of changes in the vortex wavelength on the angular momentum transport has been investigated in more detail for example by Martínez-Arias *et al.* [74] and Ostilla-Mónico *et al.* [86].

In the simulations, the axial periodicity is fixed to a relatively low number ($\Gamma = 2$) and it cannot be expected to see switching between roll wavelengths. Due to the very demanding requirements in the axial direction, confirmation of these findings in numerics requires a large amount of computational resources. However, the footprint of the rolls in the mean angular momentum fields can be visualized and the μ-range, where Taylor vortices exist, can be narrowed down more precisely. These are shown in Figure 6.11 for $Re_S = 2 \times 10^4$ for three different rotation ratios μ. For pure inner cylinder rotation, while there is a structure which can be seen in the average radial and axial velocities, no footprint is observed in the average angular momentum. For $\mu = -0.08$, the rolls are seen, but now also as footprint on the average angular momentum, confirming the experimental findings. Finally, for $\mu = -0.3$, the stabilization coming from the OC is enough to distort the rolls, and to cause the average angular momentum to exceed 0.5 within the unstable inner gap region, similar to what was seen in Ostilla-Mónico *et al.* [87] for $\eta = 0.714$.

To determine the μ-range, where Taylor vortices exist, the fraction of the large-scale circulation Nu_ω^{LSC} on the total momentum transport is calculated as it was done in Ref. [15] based on the flow field decomposition $u = \bar{u} + u''$ with $\bar{u} = \langle u \rangle_{\varphi,t}$:

$$\text{Nu}_\omega^{LSC} = \underbrace{\langle r^3 \langle \bar{u}_r \bar{\omega} \rangle_{\varphi,z,t} \rangle_r / J_\omega^{lam}}_{\text{Nu}_{\omega,stress}^{LSC}} - \underbrace{\langle r^3 \nu \partial_r \langle \bar{\omega} \rangle_{\varphi,z,t} \rangle_r / J_\omega^{lam}}_{\text{Nu}_{\omega,grad}^{LSC}} \qquad (6.4)$$

The gradient fraction of Nu_ω^{LSC} is larger than zero independent of the existence of the large-scale rolls (see Figure 6.11), which is why the focus here is on the stress fraction

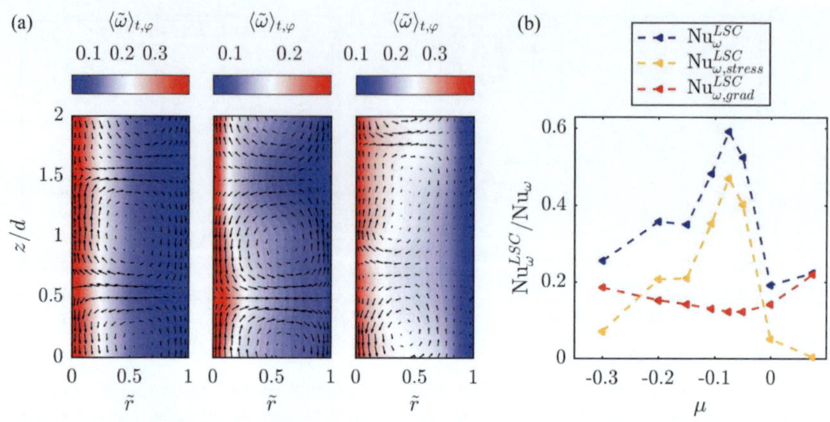

FIGURE 6.11: (a) Numerically determined azimuthally and temporally averaged angular velocity for $Re_S = 2 \times 10^4$ and $\mu = 0$ (left), $\mu = -0.08$ (middle) and $\mu = -0.3$ (right). Arrows representing the average radial and axial velocities are superimposed. (b) Contribution of the large-scale circulation to the overall Nusselt number as function of μ shown for the same driving ($Re_S = 2 \times 10^4$).

$\mathrm{Nu}_{\omega,stress}^{LSC}$. This quantity clarifies that Taylor vortices contribute considerably to the angular momentum transport in the range of $-0.3 < \mu < 0$, and in the region of the torque maximum their contribution is around 60%.

6.6   Conclusion

The angular momentum transport and the corresponding flow patterns have been investigated experimentally and numerically in a turbulent Taylor-Couette flow at a radius ratio of $\eta = 0.357$. For an outer cylinder at rest, no pure power-law scaling is found across the Re_S-range explored and the effective scaling exponent α varies noticeable in the region of $1.3 \times 10^4 \leq Re_S \leq 4 \times 10^4$. Interestingly, the frictional Reynolds number at the inner cylinder at the lower end of this region is nearly identical to the outer frictional Reynolds number at the upper end: both are approximately $Re_{\tau,crit} \approx 230$.

At first glance, this points towards a boundary layer transition that triggers the ultimate regime. However, this is ruled out by two facts: (*i*) The calculated effective scaling exponent throughout this region is smaller than the predicted lower limit of $\alpha_{marg} = 5/3$ according to King *et al.* [65] and Marcus [72], and (*ii*) non-logarithmic boundary layer profiles at the outer wall. Instead, the investigations of this study should be assigned to the classical-turbulent regime. By exploring the bulk properties, it is found that the transition itself is associated with the capacity of the outer cylinder to emit small-scale plumes. It ends as the angular momentum profile in the bulk reaches the condition of marginal stability, i.e. a flat profile at approximately the arithmetic mean $\tilde{L} = 0.5$ of both cylinders.

In the case of independently rotating cylinders, a maximum in torque was found at $\mu_{\max}(\eta = 0.357) = -0.123 \pm 0.030$, which is induced by the formation and strengthening of large-scale vortices. This value of $\mu_{\max}$ and the mechanism are in line with Ref. [15].

The state-of-the-art results of the torque maximum locations for medium and wide gaps are summarized in Figure 6.12 together with the results of this study. The good agreement with the prediction of Brauckmann and Eckhardt [15] demonstrates, that the physical mechanism for $\mu_{\max}$ does not change in that η-regime. Indeed, it could be shown that as long as radial partitioning does not appear, the Nusselt number for $\eta = 0.357$ was in good agreement to those found for $\eta \geq 0.5$, as proposed by [16]. Finally, two different wavelengths of large-scale vortices have been discovered for $Re_S > Re_{S,crit,1}$ at different forcings. In order to explore the characteristics of such multiple flow states in wide-gap TC flows in more detail, further investigations in the spirit of those of Refs. [57, 127] carried out at smaller gaps are needed.

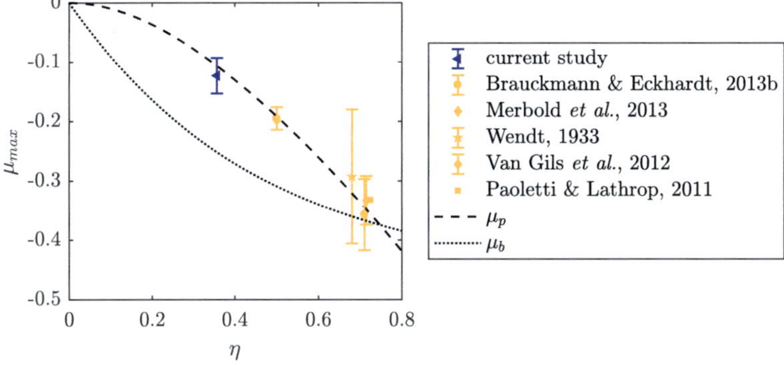

FIGURE 6.12: Torque maximum rotation rate $\mu_{\max}$ as a function of η, restricted to medium and wide gaps with $\eta \leq 0.8$. The dashed and dotted lines represent the predictions of the torque maximum according to Brauckmann and Eckhardt [15] and van Gils *et al.* [130], respectively.

Chapter 7

Characteristics of the mean velocity field in the region of the torque maximum at $\eta = 0.5$

Based on Chapter 6, it could be shown, that the physical mechanism for a maximum in torque, namely the strengthening of large-scale vortices, is the same for wide-gap TC flows of different radius ratios η ($\eta = 0.357, 0.5$), and curvature effects the momentum transport only for $\mu < \mu_{max}$. Therefore, it can be expected that the velocity field for both η features comparable characteristics in the torque maximum region. To uncover these characteristics, the velocity field is investigated for $\eta = 0.5$ using particle image velocimetry in horizontal planes at different cylinder heights for various rotation rates ($-0.6 \leq \mu \leq 0.2$), including the rotation ratio of the torque maximum $\mu_{max} = -0.2$. Additionally, numerous shear Reynolds numbers ($4 \times 10^4 \leq Re_S \leq 2.7 \times 10^5$) are analyzed in terms of the momentum transport, where Nu_ω is calculated based on the radial and azimuthal velocity components instead of direct torque measurements. The Chapter is organized as follows. In Section 7.1, the experimental setup, the investigated parameter space and measurement uncertainties are discussed. Further in Section 7.2 the averaged flow field with momentum transport relevant quantities like the radial location of the neutral line are analyzed. Afterward, the global response parameters, namely the wind Reynolds number and the Nusselt number, are evaluated in Section 7.3, to quantify especially in the latter case the contribution of the large-scale Taylor vortices to the overall momentum transport. Finally, the findings of this Chapter are summarized in Section 7.4. Note, that the results of this Chapter have been published in the Journal of Fluid Mechanics and the text is mainly taken from that paper: A. Froitzheim, S. Merbold and C. Egbers, *Velocity profiles, flow structures and scalings in a wide-gap turbulent Taylor-Couette flow*, J. Fluid Mech. 831 (2017), 330-357 [45].

7.1 Experimental setup and measurement procedure

The used setup is illustrated in Figure 7.1. The outer cylinder and the top plate are made of acrylic glass to enable optical access to the flow and both cylinders can rotate independently with maximal rotation speeds of $n_1 = 2200\,\text{rpm}$ and $n_2 = 500\,\text{rpm}$, respectively. The cylinder radii are $r_1 = 35 \pm 0.2\,\text{mm}$ and $r_2 = 70 \pm 0.2\,\text{mm}$, and the gap is enclosed by two end plates fixed at the outer cylinder leading, to a length of $\ell = 700 \pm 2\,\text{mm}$. Thus, the dimensionless geometrical parameters are $\eta = 0.5$ and $\Gamma = 20$. As working fluid distilled water is used with a kinematic viscosity of $\nu(20°C) \approx 10^{-6}\,\text{m}^2/\text{s}$, and hollow glass micro

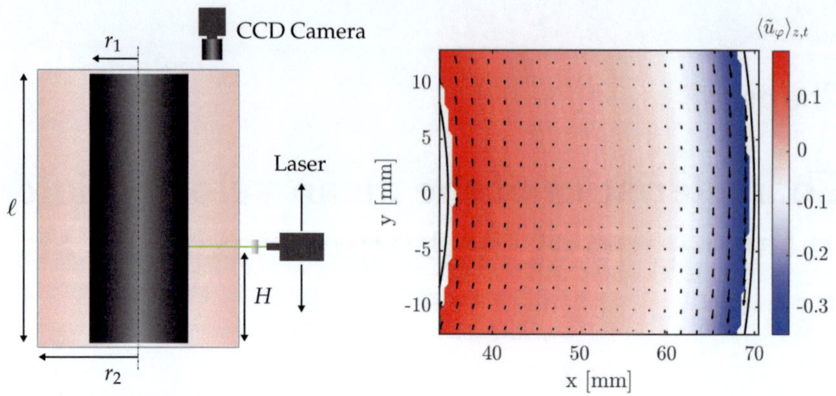

FIGURE 7.1: Sketch of the PIV setup and the temporally and axially averaged azimuthal flow field in a horizontal plane at a shear Reynolds number of $Re_S = 10^5$ and a rotation ratio of $\mu = -0.4$. For the sake of clarity only every fourth velocity vector is shown in both coordinate directions.

spheres are added as tracer particles with a mean diameter of $10 - 20\,\mu$m. The Stokes number reached within these experiments does not exceed $St = \tau_p / \tau_f < 4 \times 10^{-3}$, therefore the tracers faithfully follow the flow. The PIV setup is designed to measure the radial and azimuthal velocity components (u_r, u_φ) in a horizontal plane at different heights H (see Figure 7.1). A Flow Sense EO 2M CCD camera ($1600 \times 1200\,$px) is mounted above the transparent top plate and a Nd-YAG laser ($\lambda_{Laser} = 532\,$nm, $15\,$Hz double-frame mode) beside the TC apparatus creating a light sheet of $2\,$mm thickness. A traverse shifts the laser in a height interval of $0.571 \leq H/\ell \leq 0.686$, respectively one vortex pair is expected to be captured inside this area ($\Delta H/d = 2.29$). The main spatial properties for the outer planes are summarized in table 7.1. Further, the height dependence of the scale factor is shown in Figure 7.2(b), demonstrating the accurate alignment of the camera and the laser sheet as all dots lie on a line. Therefore, a similarity transformation between image and real coordinates can be assumed, according to Section 5.3.3.

H/ℓ	field of view [mm²]	scale factor [mm/px]	spatial resolution [mm]	OC sector [deg.]
0.571	42.1×30.6	0.0286	0.460	25.46
0.686	34.8×25.3	0.0236	0.379	20.92

TABLE 7.1: Spatial properties of the PIV setup at the top and bottom horizontal plane.

The double-frame particle images are analyzed with *Dantec Studio v. 3.31* using an adaptive PIV algorithm, which iteratively optimizes the size and shape of each IA, adapting for local flow gradients and seeding densities without the necessity of defining a prior overlap of the IA. This Multigrid-Analysis leads to a final vector spacing of $16\,$px. Further, the flow fields are post-processed using *MATLAB R2014b*.

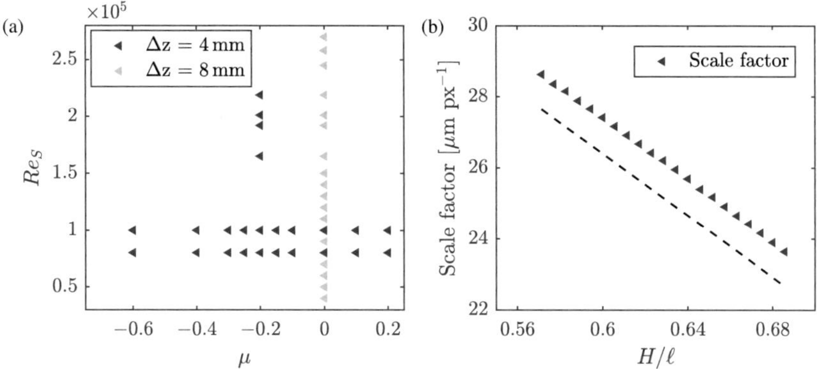

FIGURE 7.2: (a) Investigated parameter space as a function of the shear Reynolds number Re_S and the rotation ratio μ. Blue triangles represent measurements over an axial height of 80 mm with a spacing of 4 mm; yellow triangles with a spacing of 8 mm. (b) Scale factor at the 21 height positions. Dashed line serves as guide for the eye.

The parameter space investigated within this chapter is shown in Figure 7.2(a). At $Re_S = 8 \times 10^4$ and $Re_S = 10^5$ measurements are done for rotation ratios between $\mu = +0.2$ and $\mu = -0.6$. Additionally, at the torque maximum ($\mu_{max} = -0.2$) and for an outer cylinder at rest ($\mu = 0$), the flow is analyzed in the range of $1.65 \times 10^5 \le Re_S \le 2.01 \times 10^5$ and $4 \times 10^4 \le Re_S \le 2.7 \times 10^5$, respectively. The flow states are recorded at 21 axial positions with a relative displacement of $\Delta z = 4$ mm. Due to an observed less axial dependence of the flow in the case of an outer cylinder at rest, which will be shown in section 7.2.1, the axial spacing was doubled to $\Delta z = 8$ mm for $\mu = 0$. To guarantee reproducibility of the measurements, a fixed protocol was used to adjust all investigated flow states. Starting at a resting flow and room temperature of 20°C, the IC is driven up slowly to its final speed before the OC is accelerated. Then, a waiting time of around 10 min is used to ensure a stable flow state. During this waiting time, the interval Δt between the laser pulses is adjusted to enclose the particle displacement between 4 and 8 pixels, leading to 70 μs $\le \Delta t \le$ 870μs for the investigated parameter space. The measurements start at the highest cylinder position and end up at the lowest one. After each measurement the traverse with the light sheet is shifted 4 mm or 8 mm downwards and the camera focus is adjusted to the new position. Each measurement consists of 150 double-images recorded at 15 Hz, leading to a measurement time of 10 s for each height. In the case of an outer cylinder at rest, this corresponds to 39 up to 263 rotations of the IC, and for $Re_S = 10^5$ at different μ between 50 and 100 IC rotations and between 9 and 36 OC rotations. During one run including 21 height steps the increase of the fluid temperature due to friction did not exceed 1°C.

7.1.1 Measurement uncertainties

Within this study different uncertainties are concerned to quantify the accuracy of the present measurements. As a subpixel interpolation scheme is used for the PIV algorithm to detect the correlation peak, the resolvable displacement for a single measurement is limited to $\Delta_{SM} = 0.1$ px according to Nobach and Bodenschatz [81] (see Section 5.3.4). In table 7.2 the errors of the velocity measurements are exemplified for $Re_S = 10^5$. The error of a single measurement does not exceed 0.84% of the shear velocity and the standard error of the mean over t and φ not even 0.008%.

μ	u_S [m/s]	Δ_{SM} [mm/s]	$\Delta_{Mean(t,\varphi)}$ [mm/s]
$+0.2$	2.85	$19.7 - 23.9$	$0.20 - 0.22$
0	2.85	$7.9 - 9.6$	$0.08 - 0.09$
-0.2	2.85	$3.9 - 4.8$	0.04
-0.4	2.85	$3.0 - 3.6$	0.03

TABLE 7.2: Measurement uncertainty range for the 21 height positions due to a limited resolvable particle displacement detection for $Re_S = 10^5$ and different μ.

A second important source of uncertainty results from the small number of snapshots of 150 image pairs or rather the short measurement time of 10 s. It has been chosen that small because the temperature of the experiment is not actively controllable and the vortices have to be stable for the 21 height measurements performed during one run. However, turbulent flows are investigated, which means that a large number of snapshots is needed to get statistically converged results. This convergence can be quantified based on the Nusselt number computed from the velocity fields according to Section 2.5 for the highest shear Reynolds numbers investigated at $\mu = 0$ and $\mu = -0.2$ by calculating a progressing average over time (see equation (5.1)). According to Figure 7.3 the Nusselt number converges around 8 s. The remaining statistical uncertainty is calculated by taking the standard deviation over the last two seconds related to its mean, reaching 0.24% and 0.12% for the exemplified cases.

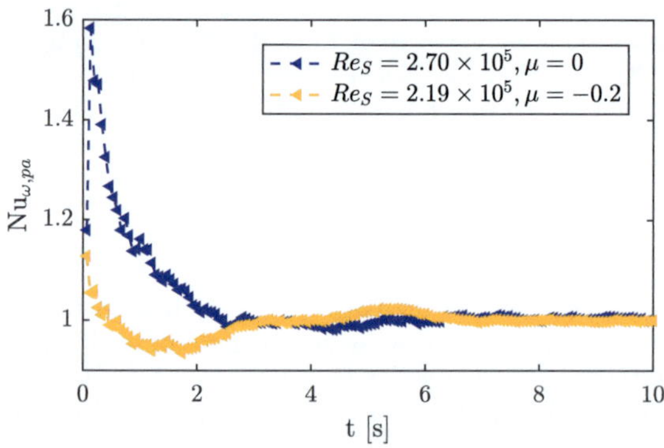

FIGURE 7.3: Progressing average of the Nusselt number as a function of time t for the highest shear Reynolds number of $Re_S = 2.70 \times 10^5$ at $\mu = 0$ and $Re_S = 2.19 \times 10^5$ at $\mu = -0.2$, respectively. Convergence is reached after a measurement time of approximately 8 s.

Further, in Figure 7.4 the time signal of the azimuthally averaged radial velocity component in the vortex inflow, vortex center and vortex outflow region at the middle of the gap together with the corresponding temporally averaged axial profile for $Re_S = 10^5$ and $\mu = -0.2$ are depicted. Obviously, the vortices at the torque maximum are stable for the measurement time and result in an axial profile, which deviates from a sinusoidal shape especially in the outflow region, in contrast to laminar Taylor vortices. Therefore, the choice of 150 snapshots taken over $10\,\text{s}$ is a suitable trade-off between statistical convergence, temperature increase and vortex stability.

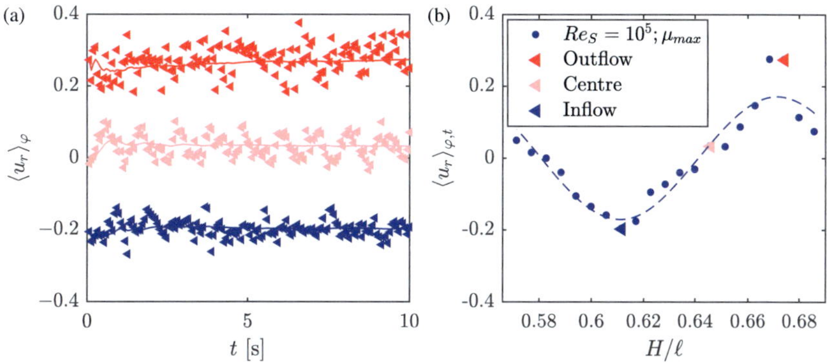

FIGURE 7.4: (a) Proof of vortex stability in terms of the azimuthally averaged radial velocity component u_r at $\tilde{r} = 0.5$ at the axial height of the vortex inflow (blue), center (pink) and outflow (red). Lines represent the corresponding temporal progressing average. (b) Azimuthally and temporally averaged axial profile of the radial velocity component with highlighted vortex inflow, center and outflow. Dashed line is a fitted sinusoidal function as guide for the eye.

Finally, also axially averaged flow properties are calculated within this paper, which requires an exact average length of one vortex pair. According to Figure 7.4, the height scan consisting of 21 equally spaced axial positions covers nearly the length of a vortex pair up to approximately one height step missing. The corresponding error becomes largest at the rotation ratio of the torque maximum, where the axial dependence of the flow is most pronounced. Therefore, this error is quantified for $\mu = -0.2$, assuming that the adjacent vortices of the measurement volume have the same shape as the measured one. Thus, simply an additional, hypothetical height position can be added below the measurement volume with the same flow state as for the top axial position to close the vortex pair. Exemplified on the axial profile of Figure 7.4(b), a u_r-value is added at $H/\ell = 0.691$ identical to the one at $H/\ell = 0.571$. The calculated deviation from the original result yields an estimation for this uncertainty. Within the following sections, all three measurement uncertainties are taken into account by the root mean square of the individual ones. Most errorbars are within the size of the used symbols and therefore not shown.

7.2 Characteristics of the flow field

7.2.1 Flow states

The Taylor-Couette flow, where both cylinders rotate independently, can be distinguished into three regimes concerning the rotation ratio μ, namely co-rotation ($\mu > 0$), low counter-rotation ($\mu_{max} \leq \mu < 0$) and high counter-rotation ($\mu < \mu_{max}$). The flow for an outer cylinder at rest is between the first two mentioned regimes. In Figure 7.5 different flow regimes are illustrated for a shear Reynolds number of $Re_S = 10^5$ and the rotation ratios $\mu = 0$, $\mu = -0.2$ and $\mu = -0.6$.

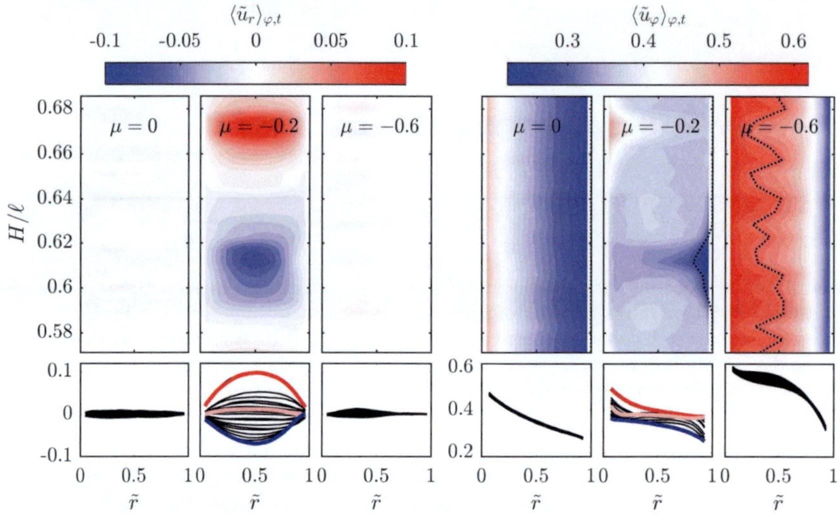

FIGURE 7.5: Flow states in terms of contour plots and velocity profiles of the azimuthally and temporally averaged, normalized radial $\tilde{u}_r$ und azimuthal velocity component $\tilde{u}_\varphi$ at 21 different heights for $Re_S = 10^5$ and the rotation ratios $\mu = 0, -0.2, -0.6$. Profiles for $\mu = -0.2$ at the axial position of the vortex inflow (blue), outflow (red) and center (pink) of a turbulent Taylor vortex are highlighted. The dotted line indicates the position of the neutral surface $\tilde{r}_n$ (see Section 7.2.3). The representation of the colormaps of the contour plots is based on a bilinear interpolation.

On the bottom part of the figure, the azimuthally and temporally averaged (φ, t) radial and azimuthal velocity profiles for 21 height positions are depicted and compounded in a contour plot above. As a reminder, the used normalization are according to Section 5.3.2. For pure inner cylinder rotation ($\mu = 0$) the temporally and azimuthally averaged radial velocity is close to zero all over the gap, with only weak strips of positive and negative velocity alternating in the axial direction. These strips can be interpreted as remnants of turbulent Taylor vortices inside the flow. Therefore, the azimuthal velocity component changes only noticeable with the radial coordinate direction and the corresponding height profiles collapse to one line. When the rotation ratio is decreased to $\mu = -0.2$, strong vortices are present in the flow and fill the whole gap. In the in- and outflow regions, angular velocity is transported strongly from the cylinders in wall-normal direction, causing an

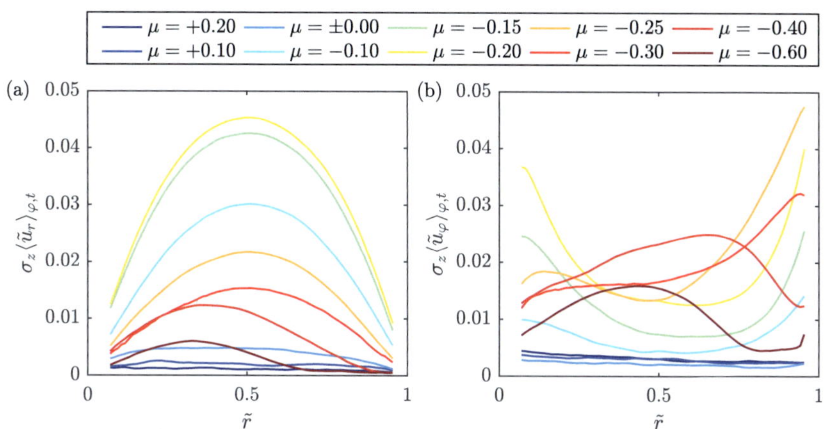

FIGURE 7.6: Standard deviation of the azimuthally and temporally averaged (a) radial and (b) azimuthal velocity component at each radial position over the 21 heights at the shear Reynolds number $Re_S = 10^5$ and different rotation ratios μ.

enhanced angular momentum transport. The evolution of these roll structures generates the maximum in torque. Further, the velocity profiles vary clearly with the axial position. The radial velocity profiles are distributed quite symmetrically about the zero line, while the azimuthal profiles do surround a nearly horizontal line with a nonzero amount. This flat azimuthal profile corresponds to the axial position of the vortex center. When μ is decreased to $\mu = -0.6$, the flow changes again significantly. The axial dependence of the profiles is now restricted to an inner part of the gap and the roll structures exist only in that inner region. In the outer gap region the flow is stabilized by the outer cylinder rotation leading back to a mainly radial dependence of the azimuthal velocity. These results are in good agreement with the findings of Chapter 6 for $\eta = 0.357$, but enable deeper insights into the flow than the local flow visualizations. Further, also the findings of van der Veen *et al.* [128] at the same radius ratio and these of Ostilla-Mónico *et al.* [83] at $\eta = 0.714$ conform with the presented results.

To analyze the evolution of the axial dependency with a decreasing rotation ratio in more detail, further the standard deviation of the azimuthally and temporally averaged velocity components is calculated at each radial position across the 21 heights for $Re_S = 10^5$. The result is depicted in Figure 7.6. For both co-rotating flow states, the radial and azimuthal velocity profiles nearly do not deviate in the axial coordinate direction. At pure inner cylinder rotation, axial dependence starts to develop in the radial velocity component with a larger amount near the inner cylinder, while the azimuthal velocity profiles still do not change. In the low counter-rotating regime, turbulent Taylor vortices are formed with an increasing strength, which is reflected by an increase of axial dependence in both profiles. The $\sigma_z \langle u_r(z) \rangle_{\varphi,t}$-profiles are symmetrically arranged around the center of the gap with a growing maximum at $\tilde{r} = 0.5$. This maximum decreases for μ beyond μ_{max} and symmetry breaks at $\mu = -0.4$. Here, the axial dependence of the profiles is restricted to the inner part of the gap and shifted away from the outer cylinder wall with decreasing μ. Accordingly, the standard deviation of the azimuthal velocity profiles in the axial direction increases first all over the gap, but it becomes more pronounced in the cylinder wall regions. For stronger counter-flow again the effect of the outer cylinder stabilization is seen with a local maximum closer to the outer cylinder wall than in the case of the radial velocity component. This evolution of the height dependence of the velocity profiles clearly shows that the turbulent Taylor vortices are formed and strengthened in a continuous process while their break down in the outer gap region seems to happen suddenly.

7.2.2 Velocity profiles

To better understand the influence of the rotation ratio μ and the shear Reynolds number Re_S on the flow, the spatially and temporally averaged (φ, z, t) radial profiles of the angular velocity $\tilde{\omega}$ and angular momentum $\tilde{L}$ with $L = r^2 \omega$ are analyzed. They can be compared to the laminar Couette solution, which has been introduced in equation (2.25) and becomes independent of μ, when normalized as shown in equation (2.26).

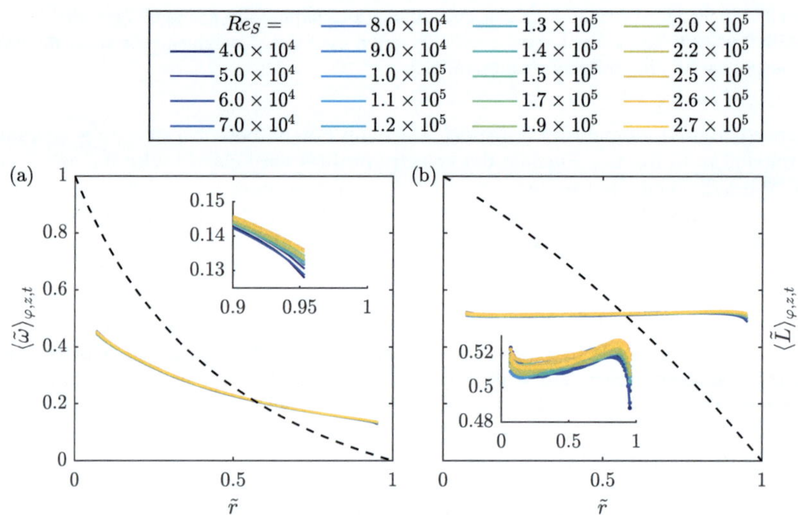

FIGURE 7.7: Spatially and temporally averaged (φ, z, t) radial profiles of the normalized angular velocity $\tilde{\omega}$ (a) and angular momentum $\tilde{L}$ (b) for pure inner cylinder rotation and different shear Reynolds numbers. Dashed lines represent the corresponding laminar profile.

Figure 7.7 illustrates the influence of the shear Reynolds number in the case of $\mu = 0$. The normalized profiles collapse to one line in the bulk flow, indicating a universal flow behavior at infinite cylinder speed. Only near the cylinder walls is a deviation visible, reflecting the decrease of the boundary layer thickness with an increasing Reynolds number. Specifically, the L-profiles are nearly flat with a slightly positive slope in the bulk and an amount of about $\tilde{L} \approx 0.5$, the arithmetic mean of the angular momenta of the cylinder walls. The same result was found by Brauckmann *et al.* [17] and linked to neutral stability of the profiles. Further, a value of $\tilde{L} \approx 0.5$ in the bulk flow is also present for

$\eta = 0.357$ after the transition found in the scaling exponent α of Nu_ω with Re_S around $Re_S = 4 \times 10^4$.

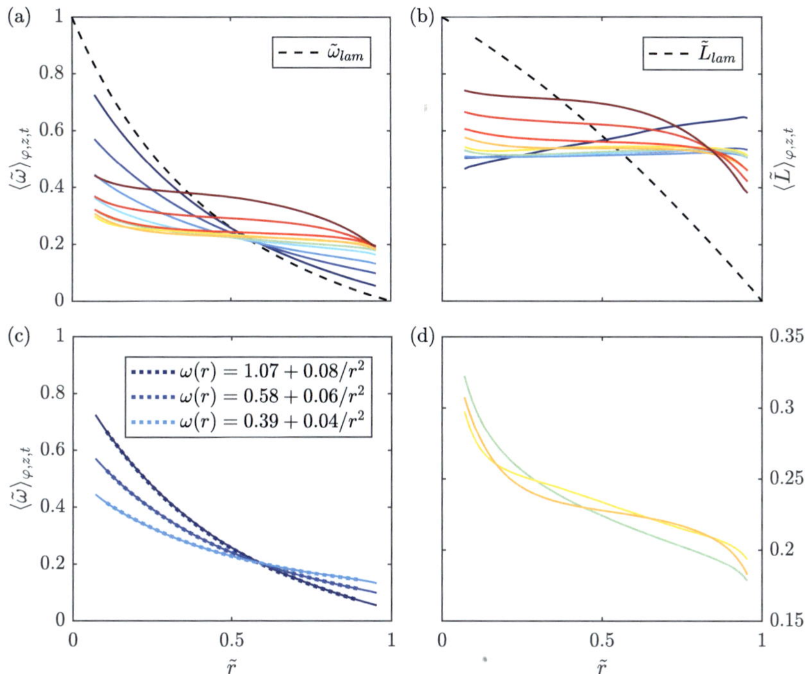

FIGURE 7.8: Spatially and temporally averaged (φ, z, t) radial profiles of the normalized angular velocity $\tilde{\omega}$ (a,c,d) and angular momentum $\tilde{L}$ (b) for $Re_S = 10^5$ and different μ. Colors according to legend of figure 7.6. Dotted lines in c) represent bulk profiles based on the laminar Couette solution with boundaries $\omega_{1,i}(\tilde{r} = 0.1)$ and $\omega_{2,i}(\tilde{r} = 0.9)$. d) Zoom in for $\tilde{\omega}$-profiles for $-0.15 \le \mu \le -0.25$.

In Figure 7.8(a) the averaged angular velocity profiles are depicted together with the laminar one at $Re_S = 10^5$ for different rotation ratios. When μ is decreased from $+0.2$ to -0.2, the profiles at first follow the laminar one quite well and become then continuously flatter until they are nearly horizontal at $\mu = \mu_{max}$. For higher counter-rotation rates beyond the maximum regime, they are shifted in the inner bulk region to higher values and adapt in the outer bulk region more and more to the laminar profile. Therefore, the torque maximum seems to coincide with the smallest gradient of angular velocity in the bulk flow, reflecting an effective mixing. Similar investigations at a much lower shear Reynolds number of around $Re_S = 3950$ have been done by Ostilla-Mónico *et al.* [87], finding the same change of profile shape. Their profiles show in addition the expected difference in boundary layer thickness at the inner and outer cylinder, as they reach far inside the gap at low Reynolds numbers.

In Figure 7.8(b) the same profiles are shown in terms of the angular momentum. For co- and weak counter-rotating flow states, the profiles are flat in the bulk with an amount of $\tilde{L} \approx 0.5$ except for $\mu = +0.2$. There, the profile exhibits a strong positive slope, which is a consequence of the normalization process as $L_1 - L_2$ goes to zero when μ approaches to the Rayleigh stability line of $\mu = \eta^2 = 0.25$. For higher counter-rotation, the L-profiles are shifted upwards in the inner bulk region, while the slope becomes more and more negative in the outer one. The sensitivity of the shape of the L-profiles to changes in μ in the high counter-rotating regime is more pronounced than for the ω-profiles and the opposite is valid for $\mu > \mu_{max}$.

Re_S	μ	$\tilde{\omega}_{1,i}$	$\tilde{\omega}_{2,i}$	μ_i	η_i	A_{lam}	B_{lam}
10^5	0.2	0.67	0.07	0.3520	0.58	1.05	0.08
10^5	0.1	0.53	0.11	0.3493	0.58	0.57	0.06
10^5	0	0.42	0.14	0.3486	0.58	0.34	0.04
8×10^4	0.2	0.67	0.07	0.3528	0.58	0.89	0.07
8×10^4	0.1	0.53	0.11	0.3490	0.58	0.44	0.05
8×10^4	0	0.42	0.15	0.3452	0.58	0.17	0.03

TABLE 7.3: Parameters of the hypothetical laminar Couette profile in the bulk TC system. $\tilde{\omega}_{1,i}$ and $\tilde{\omega}_{2,i}$ represent the values of angular velocity at $\tilde{r} = 0.1$ and $\tilde{r} = 0.9$, respectively. μ_i, η_i, A_{lam} and B_{lam} are the corresponding rotation ratio, radius ratio and coefficients of equation (2.25) of the bulk system: $\omega_{lam} = A_{lam} + B_{lam}/r^2$.

Before the average radial gradient of angular velocity and angular momentum is analyzed in the bulk flow, a closer look is taken on the profiles shape of the angular velocity for the co- and low counter-rotating regime. For $\mu \geq 0$ the flow can be separated into three regions with free surfaces in between: inner and outer boundary layer (BL) and the bulk in the spirit of the marginal stability analysis done by King *et al.* [65] and Marcus [72]. Of course, the question arises where the BLs end and the bulk starts. In general in TC systems with small radius ratios, a considerable difference between the inner ($\lambda_{BL,1}$) and outer BL thickness ($\lambda_{BL,2}$) is expected. An estimation for this difference proposed by Eckhardt *et al.* [32] is based on the assumption that the mean angular velocity in the bulk is equal to the average velocity of the cylinders. This would lead to $\lambda_{BL,1}/\lambda_{BL,2} \approx \eta^3 = 1/8$. However, this assumption does not fit to the results of Ostilla-Mónico *et al.* [87], where the BL size difference was restricted to a factor of less than $1/3$ for the present μ-regime. Further, the measurements presented here have been done at very high shear Reynolds numbers, where the size of the BLs is smaller than the closest measurement point to the wall. So, the proposed bulk system is fixed to the borders of $\tilde{r}_{1,i} = 0.1$ and $\tilde{r}_{2,i} = 0.9$, keeping in mind that the possible asymmetry of BL thickness is not inside the field of view. These borders will also be used in the following sections whenever a radial average across the bulk flow is needed. With the wall velocities of the bulk system coinciding with the locally measured angular velocities at $\tilde{r}_{1,i} = 0.1$ and $\tilde{r}_{2,i} = 0.9$, the hypothetical laminar Couette solution is calculated in the bulk system, where a nearly perfect agreement is found with the measured velocity profiles for $\mu \geq 0$ (see Figure 7.8(c)). Therefore, the angular velocity in the bulk flow is inversely proportional to the square of the radial coordinate ($1/r^2$). In table 7.3 the main parameters of the fitted laminar Couette profile in the bulk system are summarized. When the shear Reynolds number is kept constant and the rotation ratio μ is decreased from $+0.2$ to 0, the parameters A_{lam} and B_{lam} according to equation (2.25) with the corresponding bulk system values are positive and decreasing. As A_{lam} is positive, the Rayleigh stability condition $\mu_i > \eta_i^2$ is fulfilled, but the flow state comes closer to the stability border $\mu_i = \eta_i^2$ with decreasing μ. Further, a decreasing value of B_{lam} represents a decreasing magnitude of the angular velocity gradient at a specific radial position r. This shape similarity between the laminar Couette solution and fully turbulent bulk profiles of angular velocity, not only for pure inner cylinder rotation but also for co-rotation for $\eta = 0.5$, conforms with the finding of Brauckmann *et al.* [17], who reported that ω_{lam} can be used as approximation for the magnitude of the turbulent mean profile in the center of the gap for $\eta = 0.9$ and $\eta = 0.99$ and various R_Ω.

In Figure 7.8(d) also the three ω-profiles close to the torque maximum are shown to further quantify the gradient differences. In the case of $\mu = -0.15$ the profile shows a nearly linear shape in the center and outer bulk section. At the rotation ratio of the torque maximum, this linear section is enlarged into the region of the inner cylinder and the amount of the slope decreases. The profile shape changes again clearly at $\mu = -0.25$, with the local gradient decreasing further in the gap, while in the inner and outer bulk region, the gradient increases strongly. In conclusion, the locally smallest amount of angular velocity gradient is reached beyond the torque maximum rotation ratio, while at $\mu = \mu_{max}$ the profile has a long region of a small constant gradient. To compare now the ω-gradients in the bulk flow for different μ the local values are averaged over the whole length of the bulk due to its strong dependence on the radial location. For consistency, the same procedure is used for the angular momentum. In Figure 7.9 the normalized radial gradient of angular velocity and the radial gradient of angular momentum averaged over $\tilde{r} \in [0.1, 0.9]$ are depicted. As the normalized form of L is not used, the strong positive slope at $\mu = +0.2$ is prevented, resulting out of the normalization procedure.

The normalized radial gradient of angular velocity is negative for all shear Reynolds numbers and rotation ratios, as the normalized profiles are mapped to an interval of 1 at the inner and 0 at the outer cylinder. The smallest amount of gradient is found at the rotation ratio of the torque maximum, and the increase of the angular velocity gradient magnitude is stronger for $\mu > \mu_{max}$ than for $\mu < \mu_{max}$. All investigated flow states match to one curve, illustrating that the normalized bulk profiles depend less on the shear Reynolds number for $Re_S \geq 4 \times 10^4$. Ostilla-Mónico *et al.* [87] found also the smallest gradient of angular velocity coinciding with the rotation ratio of μ_{max} at $\eta = 0.5$ and $Re_S \leq 3950$, while Brauckmann *et al.* [17] could confirm this coincidence only for much smaller gaps ($\eta \geq 0.8$) at $Re_S = 2 \times 10^4$. The discrepancy is probably due to the difference in Re_S, which could also explain the Re_S-dependence of the gradient in the higher counter-rotating regime reported by Ostilla-Mónico *et al.* [87]. Further, the calculation of the gradient in the mentioned studies differs from the present ones due to limitations in the present spatial resolution. However, for co- and low counter-rotation, Ostilla-Mónico *et al.* [87] as well as Brauckmann *et al.* [17] found a linear relationship between the ω-gradient and the inverse Rossby number Ro^{-1} or rather the rotation number R_Ω. For comparison, the gradient of angular velocity is also plotted against $Ro^{-1} = 2\mu d/((1 - \mu |r_1)$ in Figure 7.9(b) for co- and weak counter-rotation, identifying this linear relationship. In the case of the angular momentum, the gradient is slightly positive for $\mu > 0$ and nearly zero at μ_{max} independent of Re_S. In the higher counter-rotating regime the gradient decreases

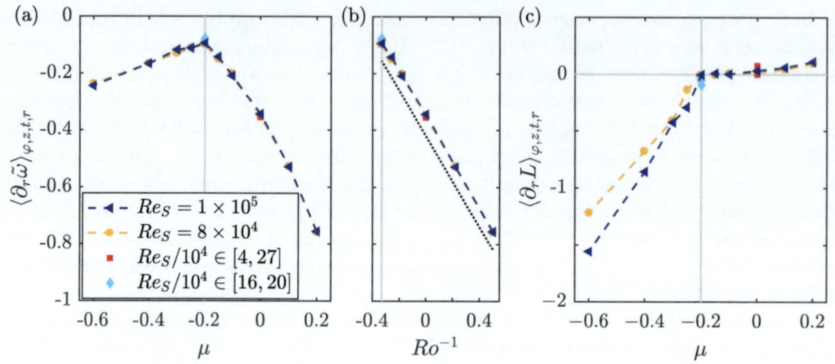

FIGURE 7.9: Normalized radial gradient of angular velocity profiles as a function of μ (a) and Ro^{-1} (b), and radial gradient of angular momentum profiles as a function of μ (c) averaged over $\tilde{r} \in [0.1; 0.9]$. Blue triangles and yellow circles represent results for $Re_S = 10^5$ and $Re_S = 8 \times 10^4$ for different rotation ratios, respectively. Red squares and cyan diamonds indicate measurements at $\mu = 0$ and $\mu = -0.2$ for different Re_S, respectively. Grey vertical lines mark the location of the torque maximum. Dotted line in (b) serves as guide for the eye.

with decreasing μ, and here the shear Reynolds number influences the slope of this decrease. However, the Reynolds number dependence vanishes when also L is normalized. Again, the bend, which coincides with μ_{max} in the present study, is shifted to smaller μ in the study of Brauckmann *et al.* [17], while the overall shape is the same.

7.2.3 Neutral surface

The velocity profiles give also access to the neutral surface r_n. It is the radial location in the gap, where the azimuthal velocity component u_φ vanishes. In the case of laminar flow this surface separates the linear stable from the linear unstable part of the flow (see equation (2.27), Chandrasekhar [21]). Once the cylinders are counter-rotating, the neutral surface detaches from the outer cylinder wall. If the linear instability sets in, the secondary flow induced by the Taylor vortices can extend this theoretical line. Esser and Grossmann [36] derived a prediction for the vortex extension called $r_{n,EG}$ (see Section 2.4.2), already expressed for TC flows by [15]:

$$r_{n,EG}(\mu) = r_1 + a(\eta)(r_{n,lam} - r_1) \quad \text{with} \quad r_{n,lam}(\mu) = r_1\sqrt{\frac{1-\mu}{\eta^2 - \mu}}. \tag{7.1}$$

The prefactor in the present configuration is $a(\eta = 0.5) = 1.5548$. When the extension of the secondary flow reaches the outer cylinder wall, intermittency evolves, causing a radial inhomogeneity inside the gap. Brauckmann and Eckhardt [15] showed that this onset of intermittency also works as a prediction for the torque maximum. By converting equation (7.1) with $r_{n,EG}(\mu_{max,EG}) = r_2$ the rotation ratio for the torque maximum is predicted at $\mu_{max,EG} = -0.191$ (see equation 3.3)). In Figure 7.10(a) the position of the neutral

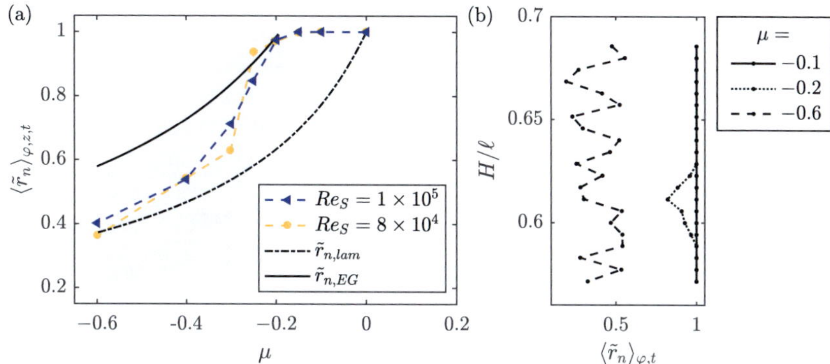

FIGURE 7.10: (a) Normalized neutral surface $\tilde{r}_n$ for $Re_S = 10^5$ (blue) and $Re_S = 8 \times 10^4$ (yellow) as a function of the rotation ratio μ. The markers (triangles, circles) indicate the averaged position over space and time (φ, z, t). The colored area represents the deviation of the neutral surface along the axial coordinate direction. (b) Radial position of the neutral surface at different heights for $Re_S = 10^5$ and $\mu = -0.1, -0.2, -0.6$.

surface for the present measurements is shown together with the two introduced predictions. The markers indicate the averaged values ($\langle \tilde{r}_n \rangle_{\varphi,z,t}$), and the colored area shows the range of this value for the different axial positions. For both Re_S the neutral surfaces are located on the outer cylinder wall or rather nested into the outer boundary layer, impossible to resolve until the rotation ratio is decreased to $\mu_{max} = -0.2$, in good agreement with the prediction of EG and LDV measurements done by van Gils *et al.* [130]. Afterward, its detachment starts first at the inflow region induced by the large-scale vortices inside the flow, leading to an axial dependency of $\tilde{r}_n$ (see Figure 7.10(b)). In the higher counter-rotating regime, the averaged measurement results and the EG prediction move apart with decreasing μ and at $\mu = -0.6$ they seem to converge with the laminar prediction. van Gils *et al.* [130] also measured the position of the neutral surface closer to the inner cylinder than the laminar prediction suggests for very strong counter-rotation, however, their measurements were only done at one cylinder height. $\tilde{r}_n$ becomes dependent on the z-coordinate for $\mu \leq \mu_{max}$ and depicts its largest range at $\mu = -0.3$, indicating the complete detachment of the neutral surface at all axial positions. Afterward, the axial range decreases again with μ, which can be explained by the vortex size. When the neutral surface is shifted from the outer cylinder into the direction of the inner one, the effective gap width of the vortices decreases, which was already shown by Ostilla *et al.* [82] for $\eta = 0.714$. Accordingly, smaller vortices lead to a smaller axial range in the location of the neutral surface. In summary, equation (7.1) predicts the start of the neutral surface detachment very well, however it fails for higher counter-rotating rates. In that regime, the laminar prediction fits very well to the data of this study, as the outer cylinder stabilization laminarizes large areas of the gap. Further, the coincidence of the $\tilde{r}_n$-detachment with the maximum of angular momentum, together with the findings of Figure 6.7 that the radius ratio only influences the momentum transport for $\mu < \mu_{max}$, demonstrates that curvature effects in TC flows become important when a radial flow partitioning occurs.

7.2.4   Energy distribution

The kinetic energy of the flow can offer a good insight into changes of the flow state concerning turbulent Taylor rolls, when the velocity field is decomposed into its large-scale and its turbulent contribution according to Brauckmann and Eckhardt [15].

$$\mathbf{u} = \langle \mathbf{u} \rangle_{\varphi,t} + \mathbf{u}'' = \bar{\mathbf{u}} + \mathbf{u}''. \tag{7.2}$$

Then, the kinetic energy averaged over space and time (φ, z, t) reads:

$$E_{kin}(r) = \frac{1}{2} \left\langle \bar{u}_\varphi^2 + \bar{u}_r^2 + \bar{u}_z^2 + u_\varphi''^2 + u_r''^2 + u_z''^2 \right\rangle_{\varphi,z,t}. \tag{7.3}$$

The mixed terms vanish and the first term in equation (7.3) reflects the azimuthal base flow energy, which is the dominant quantity. To reveal the underlying flow structure, this energy portion is neglected. Further, the axial velocity component could not be measured within the horizontal planes, so also u_z is set to zero. The remaining terms can be separated into the large-scale circulation energy E_{LSC} and the turbulent kinetic Energy E_{turb}:

$$E_{LSC}(r) = \frac{1}{2} \left\langle \bar{u}_r^2 \right\rangle_{\varphi,z,t}, \quad E_{turb}(r) = \frac{1}{2} \left\langle u_r''^2 + u_\varphi''^2 \right\rangle_{\varphi,z,t}. \tag{7.4}$$

These energies are summarized to the total kinetic energy $E_{tot}(r) = E_{turb}(r) + E_{LSC}(r)$ and normalized using the shear velocity: $\tilde{E} = E/E_S$ with $E_S = u_S^2/2$.

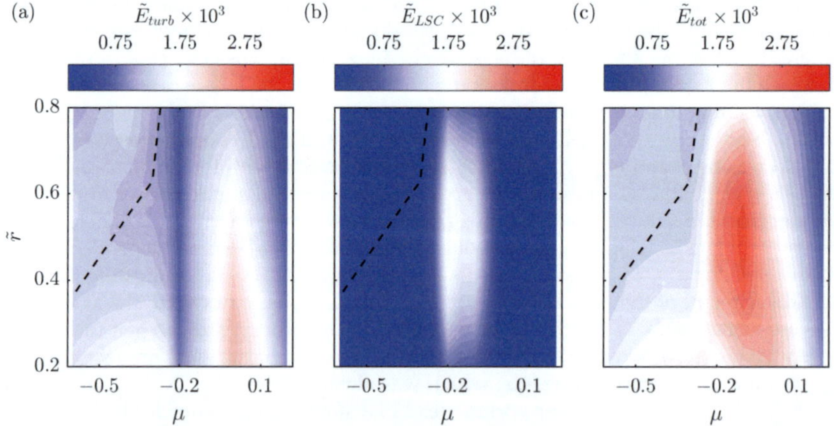

FIGURE 7.11: Contour plots of the spatially and temporally averaged (φ, z, t), normalized kinetic energy fractions at $Re_S = 8 \times 10^4$ as a function of the normalized radius $\tilde{r}$ and the rotation ratio μ. The dashed line represents the location of the neutral surface (see Figure 7.10). The representation of the colormaps of the contour plots is based on a bilinear interpolation.

In Figure 7.11 the different energy fractions are plotted as a function of $\tilde{r}$ and μ for $Re_S = 8 \times 10^4$. The turbulent kinetic energy is larger for the inner than the outer gap region and becomes dominant at pure inner cylinder rotation. In the case of co-rotation, the isolines have a curved shape, indicating a flow state close to the laminar one. At the rotation ratio of the torque maximum, the turbulent kinetic energy is reduced strongly, which illustrates a more coherent flow state induced by the formation of large-scale vortices. In the high counter-rotating regime, the neutral line is located in an area of less turbulent kinetic energy, while in the inner and outer gap region, turbulence again increases slightly. Especially near the inner cylinder a small region of stronger fluctuations is present caused by the beginning of intermittency described by Brauckmann and Eckhardt [15]. The large-scale circulation energy is close to zero at co- and high counter-rotating flow states and depicts a pronounced maximum in the low counter-rotating regime at $\mu = \mu_{max}$. While the increase of $\tilde{E}_{LSC}$ starting at $\mu = 0$ seems to be smooth, its decrease happens rapidly for $\mu < \mu_{max}$. The sum of both energy fractions results in a maximum of the total kinetic energy in the low counter-rotating regime at $\mu = -0.1$ shifted from the maximum of the large-scale circulation energy.

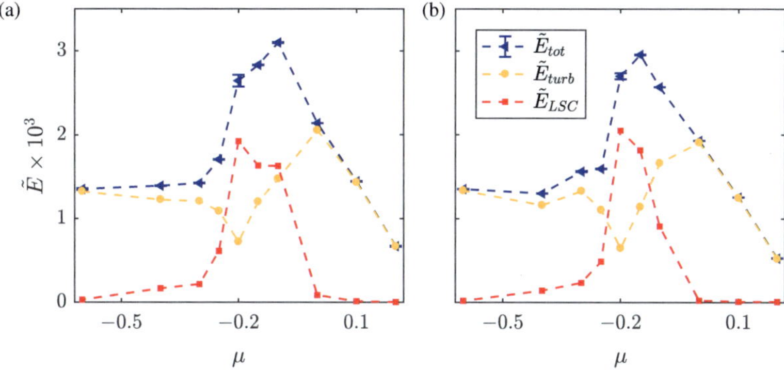

FIGURE 7.12: Spatially and temporally (φ, z, t) averaged profiles of the normalized total $(\tilde{E}_{tot})$, turbulent $(\tilde{E}_{turb})$ and large-scale circulation energy $(\tilde{E}_{LSC})$ in the center of the gap $(\tilde{r} = 0.5)$ for (a) $Re_S = 8 \times 10^4$ and (b) $Re_S = 10^5$.

In these energy calculations, only two of the three velocity components could be considered. Therefore, in Figure 7.12 the energy fractions are plotted at the center of the gap $(\tilde{r} = 0.5)$ for $Re_S = 8 \times 10^4$ and $Re_S = 10^5$. Assuming, that the axial velocity component of an idealized turbulent Taylor vortex vanishes in the center of the gap, the error of the large-scale circulation energy calculated from a 2D velocity field is reduced. Again, the large-scale circulation energy becomes maximal at the rotation ratio of the torque maximum, while the turbulent fraction exhibits a local minimum. For co- and strong counter-rotation $\tilde{E}_{LSC}$ vanishes and $\tilde{E}_{turb}$ seems to converge in the high counter-rotating regime to a constant nonzero value. When both measured Reynolds numbers are compared, a pronounced difference is seen in $\tilde{E}_{LSC}$ and $\tilde{E}_{tot}$ for $\mu = -0.1$. This difference results from a change in the vortex state for $Re_S = 8 \times 10^4$ as shown in Figure 7.13:

Despite the limited axial resolution of 4 mm, the different wavelengths of approximately $\lambda_{TV}(\mu = -0.1) \approx 0.97d$ and $\lambda_{TV}(\mu = -0.2) \approx 1.14d$ can be clearly seen based on one vortex pair. According to Ostilla-Mónico *et al.* [86] and Martínez-Arias *et al.* [74], a smaller wavelength of turbulent Taylor vortices and therefore a higher total number of vortices across the gap results in a larger transport of angular momentum in the classical

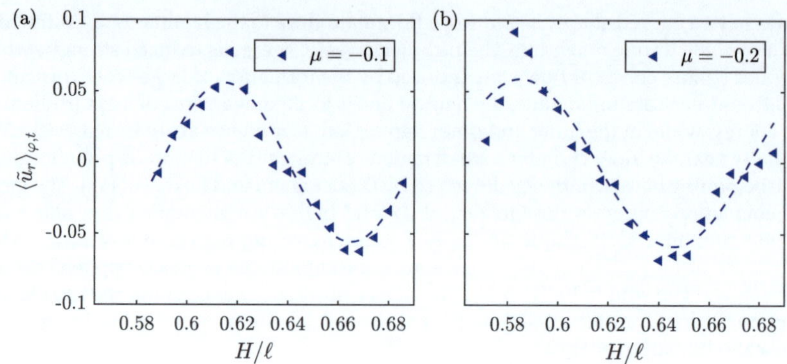

FIGURE 7.13: Axial profile of the averaged (φ, t) and normalized radial velocity component at $\tilde{r} = 0.5$ for $Re = 8 \times 10^4$ and (a) $\mu = -0.1$ and (b) $\mu = -0.2$. Dashed lines are fitted sinusoidal functions as guide for the eye.

regime and vice versa in the ultimate regime for $\eta = 0.909$. In contrast, Huisman *et al.* [57] reported a higher Nusselt number for a higher number of vortices in the ultimate regime for $\eta = 0.716$, measuring the torque only at a central segment of the inner cylinder. In the present case for $\eta = 0.5$ flow states are investigated close to the transitional point between the classical and ultimate regime (see section 7.3) and a higher total kinetic energy $\tilde{E}_{tot}$ and large-scale circulation energy $\tilde{E}_{LSC}$ is observed at a smaller value of λ_{TV}, comparing the data for $Re_S = 8 \times 10^4$ and $Re_S = 10^5$. Apparently, the present findings support the results shown by Huisman *et al.* [57]. It is also worth mentioning that the change in wavelength illustrated in Figure 7.13 is not a random event, as it was observed also by repeating the measurements. Up to this exceptional case, all flow states measured in the low counter-rotating regime within this study exhibit vortices of wavelengths of around 1.14 gap widths. Therefore, it can be concluded that the rotation ratio of the torque maximum coincides with the maximum of $\tilde{E}_{LSC}$ and a local minimum of $\tilde{E}_{turb}$, while $\tilde{E}_{tot}$ maximizes at a slightly larger μ. Further, the large-scale circulation energy decreases strongly for $\mu < \mu_{max}$, supporting the picture of a rapid breakdown of the vortices as already shown in Figure 7.6.

7.3   Global response parameters

7.3.1   Wind Reynolds number Re_W

The wind Reynolds number Re_W is a measure for the strength of the secondary flow (the so-called wind with components u_r and u_z) and its amplitude is analyzed based on the accessible radial velocity component u_r. As turbulent Taylor vortices are present inside the gap in the low counter-rotating regime, the wind amplitude is dominated by the variation of the velocity components in the axial direction. Therefore, $\sigma(u_r)$ is calculated not only over time and the azimuthal coordinate direction, but also over the axial one; the notation is $\sigma(u_r) = \sigma_{\varphi,z,t}(u_r)$. The normalization is done again by use of the shear velocity.

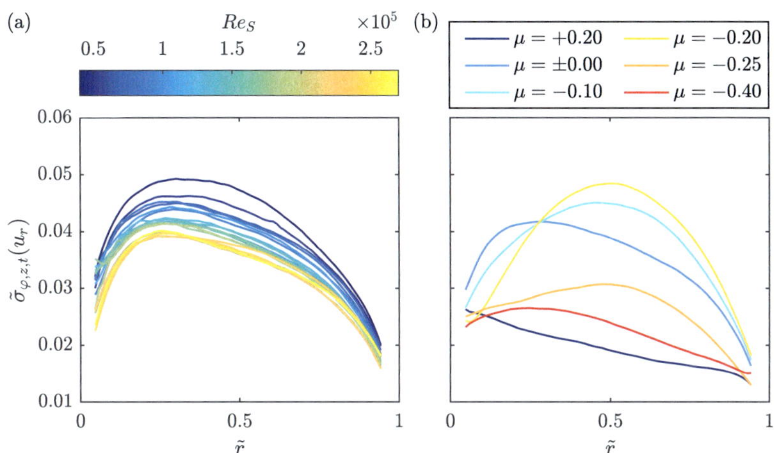

FIGURE 7.14: (a) Normalized radial profiles of the standard deviation of the radial velocity component $\tilde{\sigma}_{\varphi,z,t}(u_r) = \sigma_{\varphi,z,t}(u_r)/u_S$, calculated over space and time (φ, z, t) for different Re_S and an outer cylinder at rest ($\mu = 0$). Discrete colors for interpolated colormap according to Figure 7.7. (b) Normalized radial profiles of the standard deviation of the radial velocity component for different rotation ratios μ at $Re_S = 10^5$.

In Figure 7.14 the radial profiles of the standard deviation of the radial velocity component $\tilde{\sigma}_{\varphi,z,t}(u_r) = \sigma_{\varphi,z,t}(u_r)/u_S$ are pictured. When the outer cylinder is at rest, the profiles show a maximum at $\tilde{r} \approx 0.3$ independent of the shear Reynolds number (see Figure 7.14(a)). This asymmetric shape may be the result of the strong curvature existing in a wide-gap TC flow, which was also found by van der Veen *et al.* [127]. With increasing Re_S the profiles are shifted to smaller values while their shape stays the same. However, the shape of the profiles strongly depends on the rotation ratio μ (see Figure 7.14(b)). For $\mu = +0.2$ the profile is a nearly straight line in the bulk flow with a negative slope, meaning that higher fluctuations exist in the inner gap region. In the low counter-rotating regime, the profiles again depict a maximum, which is now located in the center of the gap and increases until $\mu = \mu_{max}$ is reached, reflecting the strong variation of the radial velocity component with the axial coordinate due to the dominant roll structures. Nevertheless, the profile is not symmetric to the center of the gap. This becomes clearer for $\mu = -0.25$ where the maximum is decreasing but still at the same radial location. At $\mu = -0.4$ the standard deviation further decreases especially in the outer gap region.

To calculate the wind Reynolds number out of these profiles, an adequate averaging over the radial coordinate has to be done. van der Veen *et al.* [127] suggested to perform this averaging over an interval centered by the maximum location of the profiles at $\mu = 0$, but they also used different methods. Considering the different profile shapes at the investigated rotation ratios, it can be concluded that an average over the whole gap is needed, only excluding the boundary layers ($0.1 \leq \tilde{r} \leq 0.9$). The results are presented in Figure 7.15 together with the data of van der Veen *et al.* [127]. For pure inner cylinder rotation ($\mu = 0$) the wind Reynolds number is increasing with the shear Reynolds number and a nearly perfect agreement is found with the measurements of van der Veen

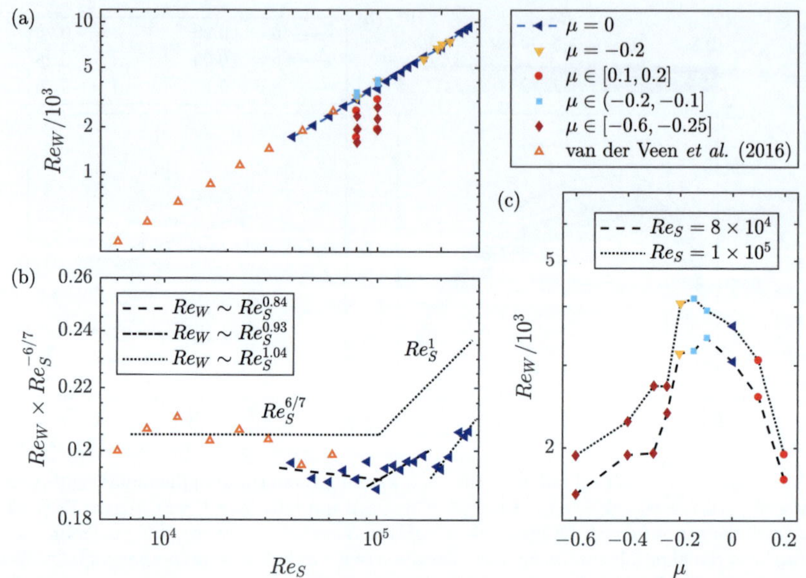

FIGURE 7.15: (a) Wind Reynolds number Re_W as a function of the shear Reynolds number Re_S. The markers and colors separate the rotation ratio regimes, namely co-rotation (red circles), outer cylinder at rest (blue triangle), low counter-rotation (cyan squares), rotation ratio of the torque maximum (yellow downwards triangles), high counter-rotation (dark red diamonds) and data taken from van der Veen *et al.* [127] for $\mu = 0$ (open upwards orange triangles). (b) Wind Reynolds number compensated with the 'classical' scaling $Re_W Re_S^{6/7}$ as a function of Re_S. Black lines represent least square fits in the classical (dashed), transitional (dash-dotted) and ultimate regime (dotted). (c) Wind Reynolds number as a function of μ for $Re_S = 8 \times 10^4$ (dashed line) and $Re_S = 10^5$ (dotted line).

et al. [127] in the lower Re_S regime. According to Grossmann and Lohse [49, 50] the wind Reynolds number scales as $Re_W \sim Re_S^{\beta}$ with $\beta = 6/7$ in the classical and $\beta = 1$ in the ultimate regime. Therefore, in Figure 7.15(b) also the wind Reynolds number compensated with the predicted scaling in the classical regime is shown. Focusing on the data of this thesis, three different scaling regimes can be assumed. The first regime is valid up to $Re_S = 10^5$ ($Ta = 1.69 \times 10^{10}$) with a scaling exponent of $\beta = 0.84 \pm 0.07$ very close to the classical $\beta = 6/7$. Secondly, some kind of transition takes place with an exponent of $\beta = 0.93 \pm 0.05$, before the last regime at $Re_S = 2 \times 10^5$ ($Ta = 6.46 \times 10^{10}$) is reached with an exponent of $\beta = 1.04 \pm 0.08$. The third one is in the range of the prediction for the ultimate regime. Although the exponents fit quite well to the expected ones, the significance of the results has to be considered carefully. Only a few partially scattering data points could be analyzed within the three mentioned regimes and the scaling seems not to be that strict, especially when also including the data of van der Veen *et al.* [127]. Further, a jump is seen between the second and third regime, indicating that ultimate turbulence is still evolving. Nevertheless, it can be stressed that the close accordance with the predictions confirms the finding of the transition from classical to ultimate turbulence. Moreover, in Figure 7.15(c) the wind Reynolds number is depicted over the rotation ratio μ. What can be observed here, represents a similar behavior to what was explained for the total kinetic energy with a maximum in the low counter-rotating regime for a rotation ratio slightly larger than μ_{max}.

7.3.2 Nusselt number Nu_ω

All investigations presented up to now were focused on the change in the flow field, when the regime of the torque maximum is passed. Now, the angular momentum itself is analyzed in terms of the Nusselt number. Therefore, next to the classical calculation according to equations (2.41) and (2.44) for the radial profiles of the pseudo-Nusselt number, also the flow field decomposition introduced in equation (7.2) is used. Thus, the Nusselt number reads:

$$\mathrm{Nu}_\omega = \mathrm{Nu}_\omega^{turb} + \mathrm{Nu}_\omega^{LSC}, \tag{7.5}$$

with its turbulent and large-scale circulation contributions

$$
\begin{aligned}
\mathrm{Nu}_\omega^{turb} &= \left\langle r^3 \left\langle u_r'' \omega'' \right\rangle_{\varphi,z,t} \right\rangle_r / J_\omega^{lam}, \\
\mathrm{Nu}_\omega^{LSC} &= \left\langle r^3 \left(\left\langle \bar{u}_r \bar{\omega} \right\rangle_{\varphi,z,t} - \nu \partial_r \left\langle \bar{\omega} \right\rangle_{\varphi,z,t} \right) \right\rangle_r / J_\omega^{lam}.
\end{aligned}
\tag{7.6}
$$

Next to the mixed terms $\left\langle \bar{u}_r \omega'' \right\rangle_{\varphi,z,t}$ and $\left\langle u_r'' \bar{\omega} \right\rangle_{\varphi,z,t}$, also $\partial_r \left\langle \omega'' \right\rangle_{\varphi,z,t}$ vanishes, as they all are linear in the deviation quantities. It remains a Reynolds stress term in the turbulent part and a gradient term and the product of the averaged velocities in the large-scale circulation part. The last one can be corrected, assuming that the radial velocity vanishes when it is averaged over the whole measurement volume. The additional radial average of the decomposition products is necessary, as they depend on the radial location (see [15]).

In Figure 7.16 and Figure 7.17 the radial profiles of the Nusselt number and the averaged value are depicted for different shear Reynolds numbers at $\mu = 0$. The Nusselt number increases with the shear Reynolds number and remains nearly constant for all radial positions, which is in good agreement with the definition of the angular momentum flux (see Figure 7.16(a)). In the region of the cylinder walls, the deviation of Nu_ω from a horizontal line increases. This becomes clearer when these profiles are divided by the radially averaged Nusselt number (see Figure 7.16(b)). Especially in the region of the inner cylinder, the profiles are below their average value, indicating a radial dependent underestimation. In the investigated Re_S-regime the Reynolds stress term $\left\langle u_r'' \omega'' \right\rangle$, which consists of the product of the velocity fluctuations, represents between 91% up to 99% of the overall angular momentum flux in the fully turbulent regime at pure inner cylinder rotation. As the spatial resolution of the velocity measurements in the different horizontal planes is limited to $0.379\,\mathrm{mm} - 0.46\,\mathrm{mm}$, these fluctuations are averaged over the corresponding grid. Further, the turbulent kinetic energy (see Figure 7.11) is very large in the region of the inner cylinder at $\mu = 0$. Therefore, the overall underestimation of the velocity fluctuations is more pronounced in the inner gap region than in the outer one.

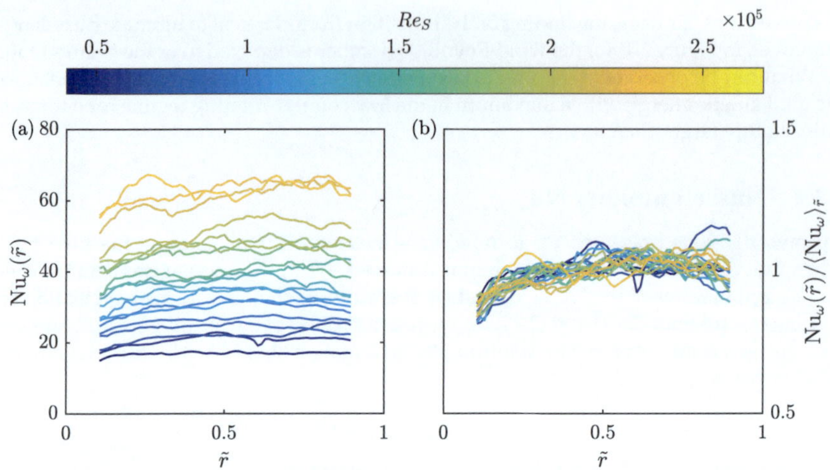

FIGURE 7.16: (a) Radial profiles of the Nusselt number $\mathrm{Nu}_\omega(\tilde{r})$ for different shear Reynolds numbers Re_S and pure inner cylinder rotation ($\mu = 0$). (b) Radial profiles of the Nusselt number divided by its average value in the range of $\tilde{r} \in [0.1; 0.9]$. Discrete colors for interpolated colormap according to Figure 7.7.

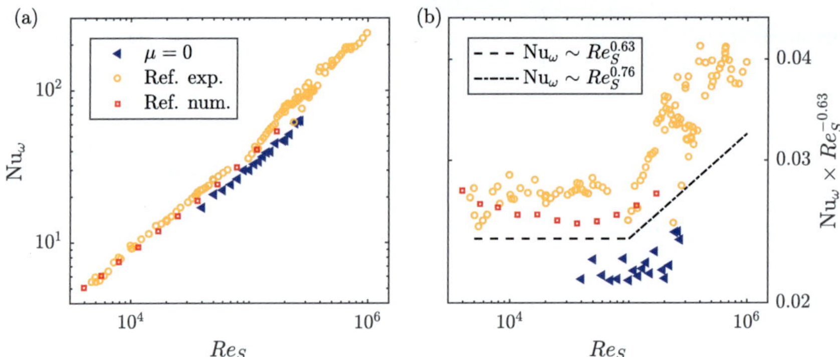

FIGURE 7.17: (a) Nusselt number Nu_ω averaged over the radial coordinate $\tilde{r} \in [0.1; 0.9]$ as a function of Re_S for $\mu = 0$. Blue filled triangles represent the values calculated out of the PIV measurements. Yellow open circles are taken from direct torque measurements [78] (Ref. exp.) and red open squares from numerical simulations [86, 87] (Ref. num.) as comparison. (b) Compensated Nusselt number $\mathrm{Nu}_\omega Re_S^{-0.63}$ as a function of Re_S. The dashed and dash-dotted lines indicate least square fits using the present PIV data for the classical and ultimate regime, separated by $Re_{S,crit} = 10^5$.

The same effect is also visible in Figure 7.17 (a), where the PIV results are compared to direct torque measurements of Merbold *et al.* [78] and direct numerical simulations of Ostilla-Mónico *et al.* [86, 87]. The absolute values of Nu_ω are underestimated for all

Re_S, but are in the same order as the comparative data. Next to the restricted spatial and temporal resolution, also the small measurement volume compared to the whole TC volume comes here into account. Although it is not possible to clearly assign the most dominant reason for the underestimation, differences induced by the wavelength of turbulent Taylor rolls can be excluded, as their influence is negligible at pure inner cylinder rotation due to their low strength (see Figure 7.5).

Ostilla-Mónico *et al.* [86, 87]		Merbold *et al.* [78]		current study	
Re_S	Nu_ω	Re_S	Nu_ω	Re_S	Nu_ω
5.38×10^4	24.1	5.05×10^4	25.6	5.02×10^4	20.7
7.90×10^4	31.3	-	-	8.00×10^4	26.3
1.16×10^5	40.9	1.20×10^5	41.3	1.20×10^5	34.2
1.70×10^5	53.9	1.66×10^5	59.8	1.66×10^5	44.9
-	-	2.20×10^5	80.1	2.19×10^5	51.5

TABLE 7.4: Comparison of Nusselt numbers for pure inner cylinder rotation between numerical simulations of Ostilla-Mónico *et al.* [86, 87], direct torque measurements of Merbold *et al.* [78] and this study.

From table 7.4 it can be seen that the differences between the data of Merbold *et al.* [78] and the present measurements increase from approximately 19% at $Re_S = 5 \times 10^4$ to 36% at $Re_S = 2.19 \times 10^5$, while those compared to Ostilla-Mónico *et al.* [86, 87] stay more or less constant around 16%. The increase in the deviation with the shear Reynolds number is what one would expect, as with higher Re_S the length- and time-scales of turbulent flows become smaller and therefore, the error due to a limited temporal and spatial resolution should increase. However, the direct torque data show partially stronger variations at specific Re_S, which complicates the quantification of the differences. Nevertheless, the scaling exponent α of the Nusselt number ($Nu_\omega \sim Re_S^{\alpha-1}$) of the present PIV data can be further compared with the corresponding direct torque measurements and numerical simulations depicted. Here, it is important to note that α depends on the shear Reynolds number Re_S, reaching a nearly constant value not before $Re_S \approx 6 \times 10^5$ [78]. Owing to the small number of measurement points, it does not make sense within this study to calculate a local exponent. However, Merbold *et al.* [78] also reported an averaged exponent before and after the transition from the classical to the ultimate regime at approximately $Re_S \approx 10^5$, which is used as a comparison. In Figure 7.17(b) the compensated Nusselt number reveals this change in scaling clearly for the direct torque data, but only blurred for the present PIV measurements and the numerical simulations. If the data of this study are fitted to a power-law ansatz, a scaling exponent of $\alpha = 1.63 \pm 0.11$ is found in the classical regime and $\alpha = 1.76 \pm 0.08$ in the ultimate regime, lying in the confidence interval of $\alpha = 1.62 \pm 0.04$ and $\alpha = 1.78 \pm 0.06$ of Merbold *et al.* [78] in both regimes. In addition, in Ostilla-Mónico *et al.* [86, 87] a scaling exponent of $\alpha \approx 1.66$ in the classical regime for $\eta = 0.5$ and an exponent of $\alpha \approx 1.76$ in the ultimate regime independent of the radius ratio is reported. This good agreement between experimental data using different measurement techniques and numerical data is quite interesting, as the differences in the Nusselt number should increase with Re_S due to resolution limitations as commented before. One explanation might be the fact that a fixed scaling exponent is not yet reached in the shear Reynolds number regime, where the PIV measurements have been performed. Also, the reported partially stronger variation in the direct torque data comes here into account.

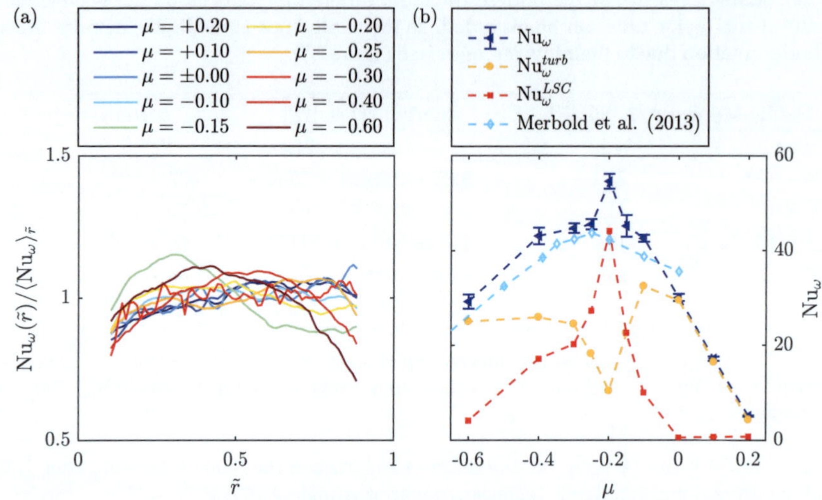

FIGURE 7.18: (a) Normalized radial profiles of the Nusselt number $\mathrm{Nu}_\omega(\tilde{r})$ for the shear Reynolds number $Re_S = 10^5$ and different rotation ratios. (b) Nusselt number Nu_ω (blue triangles) decomposed into its large-scale (red squares) and turbulent contributions (yellow circles), which are averaged over the radial coordinate, as a function of μ for $Re_S = 10^5$. Cyan open diamonds represent direct torque measurement data taken from Merbold *et al.* [78] for reference.

The dependence of the Nusselt number on the rotation ratio is analyzed based on the measurement data at $Re_S = 10^5$ using the decomposition according to equation (7.6). In Figure 7.18(a) the radial profiles of Nu_ω are depicted, showing a stronger radial deviation compared to the previous ones for $\mu = 0$. Further, in Figure 7.18(b) the different fractions of the Nusselt number are plotted against the rotation ratio μ and compared to direct torque measurements of Merbold *et al.* [78]. Similar to the results of the energy distribution, the angular momentum transport due to the large-scale circulation is around zero for $\mu \geq 0$ and increases strongly in the low counter-rotating regime with a pronounced maximum at $\mu = \mu_{max}$. Its decrease for higher counter-rotation develops slower than observed for $\tilde{E}_{LSC}$ in Figure 7.12. $\mathrm{Nu}_\omega^{turb}$ exhibits a local and sharp minimum at μ_{max} with a contribution smaller than 50 % of the overall transport and in summary the total Nusselt number maximizes at $\mu_{max} = -0.2$. In comparison to the direct torque measurements of Merbold *et al.* [78], the Nu_ω-values of this study are larger for $\mu < 0$, where the large-scale fraction cannot be neglected, while at $\mu = 0$, the opposite case is existent. So, it can be concluded, that the turbulent fraction of the Nusselt number is underestimated, while the LSC-fraction is overestimated, which is most pronounced at μ_{max}. This result enables a good insight into the evaluation of the angular momentum transport based on velocity fields. While a limited spatial resolution in the horizontal plane damps out velocity fluctuations, it over-rates the averaged secondary flow field mainly due to the more significant limited axial resolution. However, it is worth mentioning that the experimental data with which the comparison is done result from different facilities, different measurement techniques and different endplate conditions, leading to possibly different vortex wavelengths. Thus, despite differences in the Nusselt number, the overall agreement between the total Nusselt numbers is quite good. The same decomposition of the Nusselt number for $\eta = 0.5$ at a much lower Reynolds number of $Re_S = 2 \times 10^4$ has been done by Brauckmann and Eckhardt [15], finding a much smoother changeover from an angular momentum transport dominated by turbulent fluctuations for $\mu > 0$ to one dominated by large-scale circulations for $-0.6 \leq \mu \leq -0.1$ and back again. This clearly shows, that the rotation ratio range, where turbulent Taylor vortices can exist inside the gap, shrinks with increasing shear Reynolds number. When Figure 7.18(b) is compared with the visualization results of Figure 6.10(a) for $\eta = 0.357$, the analogy of the formation process of the torque maximum in both η-systems becomes obvious. It remains the question if Taylor vortices can survive in the torque maximum region for infinite forcing.

7.4 Conclusion

In this Chapter an axial exploration of a wide-gap Taylor-Couette flow is presented at high shear Reynolds numbers using PIV in horizontal planes at different cylinder heights to analyze the flow structure for co- and counter-rotation. Special attention was paid to the region of the maximum transport of angular momentum in the low counter-rotating regime ($\mu = \mu_{max}$), where turbulent Taylor vortices are formed in the gap. It could be shown, that these vortices are statistically stable in time, which makes deeper investigations possible. To verify the prediction for the torque maximum according to Brauckmann and Eckhardt [15], the angular velocity and angular momentum profiles are investigated in the bulk, finding the smallest radial gradient of angular velocity and the start of the detachment of the neutral surface from the outer cylinder wall coinciding with μ_{max}. While these findings are in good agreement with those of van Gils $et\ al.$ [130] and Ostilla-Mónico $et\ al.$ [87], they contradict the studies of Ostilla $et\ al.$ [82] and Brauckmann $et\ al.$ [17]. It can be assumed that these differences result from differences in the investigated shear Reynolds number regimes. As the results of this study are only less dependent on Re_S, it can be stressed that high shear Reynolds number investigations are needed to reveal general flow behaviors. Further, a shape similarity is detected between the laminar Couette solution and fully turbulent bulk profiles for $\mu \geq 0$, which calls for additional investigations in the future. In a next step, the velocity field is decomposed into its large-scale and turbulent contributions and the corresponding kinetic energy fractions are calculated. Here, a maximum of the large-scale circulation energy and a local minimum of the turbulent kinetic energy are found at the rotation ratio of the torque maximum. This clearly shows the increase of coherence in the flow at $\mu = \mu_{max}$. Furthermore, while the growth of turbulent Taylor vortices takes place in a continuous process, their breakdown occurs suddenly. Unexpectedly, the sum of both energy fractions maximizes at a slightly larger rotation ratio than μ_{max}. Moreover, the wind Reynolds number is calculated for different rotation ratios, showing that it is necessary to include also the axial dispersion into the standard deviation of the wind velocity. For pure inner cylinder rotation, the beginning of the transition from the classical to the ultimate turbulent regime could be identified at $Re_S = 10^5$ based on its scaling exponent in accordance with Huisman $et\ al.$ [57] and van der Veen *et al.* [127]. Interestingly, the change of Re_W with the rotation ratio is similar to the one found for the total kinetic energy with a maximum located at μ slightly larger than μ_{max}. In the end, also the Nusselt number is analyzed, finding a blurred change in the scaling exponent with the shear Reynolds number again at $Re_S = 10^5$ and values of the scaling exponent close to those of Merbold *et al.* [78] and Ostilla-Mónico *et al.* [86, 87]. In addition, the formerly introduced flow field decomposition is used, finding a maximum of Nu_ω^{LSC} and a local minimum of $\mathrm{Nu}_\omega^{turb}$ at the rotation ratio of the torque maximum. In comparison to the results of Brauckmann and Eckhardt [15], the range of rotation ratios, where turbulent Taylor vortices can exist, shrinks strongly with the shear Reynolds number. In conclusion, this study revealed some new insights into a wide-gap Taylor-Couette flow in the fully turbulent regime around the vortex-dominated torque maximum region.

Chapter 8

Small-scale statistics of the angular momentum transport and local pattern formation at $\eta = 0.714$

So far, the global angular momentum transport and the underlying flow organization concerning flow visualizations and the mean velocity field have been investigated. Within this Chapter, an experimental analysis of the influence of the large-scale Taylor rolls on the small-scale flow statistics and their interaction with small-scale plumes is provided in a fully turbulent TC flow. Therefore, a similar PIV setup as in Chapter 7 has been used in the BTTC facility at the University of Twente for a radius ratio of $\eta = 0.714$. Here, only flow states with ($\mu_{max} = -0.36$) and without ($\mu = 0$) prominent Taylor vortices are compared. The Chapter is organized as follows: In Section 8.1, the experimental setup, the measurement technique and the investigated parameter space are described in detail. Thereafter, the flow states, statistical profiles and velocity PDFs are shown and compared to the literature to prove the quality of the measurements and discuss the influence of large-scale turbulent Taylor vortices on the global flow statistics (Section 8.2). In Section 8.3 the global and local angular momentum transport are analyzed based on the net convective Nusselt number and the contributions of the vortex in- and outflow to the overall transport are worked out. To detect intermittent bursting small-scale structures which influence this transport, the PDFs of the net convective Nusselt number are evaluated over cylindrical surfaces as well as at the axial height of the vortex inflow, center and outflow in Section 8.4. The energy content and azimuthal length scale of these structures are calculated in Section 8.5 based on azimuthal energy co-spectra, while azimuthally traveling waves superimposed on the turbulent Taylor vortices are extracted from the flow field based on a complex proper orthogonal decomposition (CPOD) in Section 8.6. The Chapter ends with a summary and a conclusion (Section 8.7). Note, that the results of this Chapter have been published in the Journal of Fluid Mechanics and the text is mainly taken from that paper: A. Froitzheim, R. Ezeta, S.G. Huisman, S. Merbold, C. Sun, D. Lohse and C. Egbers, *Statistics, plumes and azimuthally traveling waves in ultimate Taylor-Couette turbulent vortices*, J. Fluid Mech. 876 (2019), 733-765 [43].

8.1 Experimental setup and measurement procedure

The PIV experiments were performed in the BTTC facility at the University of Twente. The BTTC is an ideal facility to perform PIV experiments due to its transparent outer cylinder and top plate. The inner and outer radius of the setup are $r_1 = 75\,\text{mm}$ and

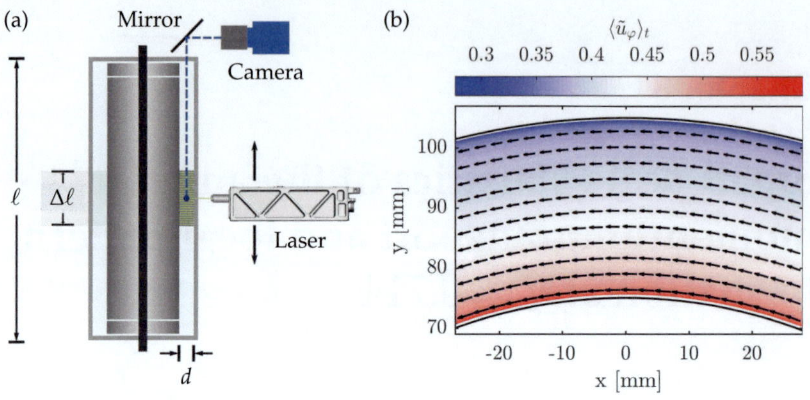

FIGURE 8.1: (a) Sketch of the BTTC apparatus in a vertical section. A mirror set at 45 degrees is used so the camera captures the velocity field in the $r - \varphi$ plane at 23 positions along $\Delta\ell$. (b) Temporally averaged flow field in a horizontal plane at the axial height of the vortex center for $Re_S = 3.51 \times 10^5$ and $\mu = 0$. Colors represent the azimuthal velocity component and arrows the velocity field. Only each tenth vector is shown. Black solid lines indicate the IC and OC positions.

$r_2 = 105\,\text{mm}$, respectively, and thus the radial gap is $d = r_2 - r_1 = 30\,\text{mm}$. The height of the cylinders is $\ell = 549\,\text{mm}$, which gives an aspect ratio of $\Gamma = \ell/d = 18.3$. The radius ratio is then $\eta = r_1/r_2 = 0.714$. A more detailed overview of the setup can be found in Huisman *et al.* [59]. The flow consists of water, whose viscosity v and density ρ can be controlled throughout the experiments due to the temperature control of the BTTC. The temperature of the experiments is fixed to $20\,^\circ\text{C}$, leading to $v = 1.002\,\text{mm}^2/\text{s}$ and $\rho = 0.998\,\text{g}/\text{cm}^3$. The standard deviation of the temperature is $15\,\text{mK}$. The flow is seeded with fluorescent polyamide particles with diameters up to $20\,\mu\text{m}$ and an average particle density on the particle images of approximately 0.01 particles/pixel. These particles are coated with Rhodamine B, which has a maximum emission centered at approximately $565\,\text{nm}$. The illumination is provided by a laser sheet from a double-pulsed cavity laser (*Quantel Evergreen* 145 laser, $\lambda_{Laser} = 532\,\text{nm}$). The thickness of the laser sheet is $\approx 1\,\text{mm}$. The laser is mounted on a traverse system (*Dantec Dynamics* lightweight traverse), which allows to precisely change the location of the laser sheet along the vertical direction. The camera used for the recordings is an Imager sCMOS ($2560 \times 2160\,\text{px}$) 16 bit with a *Carl Zeiss* Milvus 2.0/100 lens. The velocity fields are captured with a framerate of $15\,\text{Hz}$. Since the camera is operated in double-frame mode, very small interframe times are adjustable, i.e. $\Delta t \ll 1/15\text{Hz}$. In order to maximize the contrast of the images, a long-pass filter is used in front of the lens (*Edmund Optics* high-performance longpass filter, $550\,\text{nm}$), which collects only the emitted light from the fluorescent particles. In Figure 8.1(a), a sketch of the experimental setup is shown.

The velocity fields are calculated with a commercial software (*Davis 8.0*) using a multi-pass method. The algorithm uses IAs of size $64 \times 64\,\text{px}$ for the first pass and IAs of $24 \times 24\,\text{px}$ for the last iteration with a 50% overlap of the windows. This process yields the velocity fields in Cartesian coordinates. In order to have access to the velocity fields

Case	Regime	Re_S	μ	Abbreviation	Linestyle
1	classical	9.32×10^3	0	$C1_{CR,0}$	---
2	classical	9.30×10^3	-0.15	$C2_{CR,M}$	—
3	ultimate	2.98×10^4	-0.36	$C3_{UR,M}$	—
4	ultimate	6.68×10^4	-0.36	$C4_{UR,M}$	—
5	ultimate	9.46×10^4	-0.36	$C5_{UR,M}$	—
6	ultimate	2.15×10^5	0	$C6_{UR,0}$	---
7	ultimate	2.14×10^5	-0.36	$C7_{UR,M}$	—
8	ultimate	3.51×10^5	0	$C8_{UR,0}$	---

TABLE 8.1: Overview of investigated flow states, defined by the regime, the shear Reynolds number Re_S and the rotation ratio μ. The penultimate column depicts the abbreviations for the different flow states and the last column depicts the linestyles used in the study. The transition point from the classical to the ultimate regime is located at $Re_S \approx 1.6 \times 10^4$.

in polar coordinates, the Cartesian velocity fields are mapped onto a polar grid using a bilinear interpolation. The mapping is done such that the radial Δr and azimuthal $\Delta \varphi$ resolution is the same as the spatial resolution in Cartesian coordinates Δx, i.e. $\Delta r = \Delta x$ and $r\Delta \varphi = \Delta x$.

In table 8.1, a summary of the performed measurements is presented. In total, 8 cases were investigated, which will be addressed in the following sections. Each case contains measurements done at 23 different heights which are separated by $\Delta z = 4\,$mm, and each height contains 1500 different velocity fields. Thus, the height of the experiments spans a length of $\Delta \ell = 22\Delta z = 88\,$mm. The resolution of the velocity fields Δx depends on the height but lies within $\Delta x \in [0.607, 0.752]\,$mm, where the smallest value corresponds to the height closest to the camera at $(z - \ell/2)/d = 1.5$ and the smallest to $(z - \ell/2)/d = -1.5$.

Flow states are investigated for pure inner cylinder rotation as a reference and for the rotation ratio that corresponds to the torque maximum, where pronounced large-scale rolls are present in the gap (see table 8.1). Further, the measurements are classified based on the discovered change in the local scaling exponent α by Ostilla-Mónico *et al.* [83] at $Re_S(\eta = 0.714) \approx 1.6 \times 10^4$ for $\mu = 0$, which is caused by a transition of the boundary layers (BLs). Accordingly, flow states at $Re_S < 1.6 \times 10^4$ are assumed to be in the so-called classical regime with laminar BLs, while those at $Re_S > 1.6 \times 10^4$ are assumed to be in the ultimate regime with turbulent BLs. Cases 1 and 2 are in the classical regime, where, μ_{max} changes with the shear Reynolds number, which is why $\mu_{max} = -0.15$ is different to the cases in the ultimate regime [82]. For the flow states in the ultimate regime, the torque maximum is located around $\mu_{max} = -0.36$.

8.2 Flow states and velocity profiles

The TC flow at μ_{max} is dominated by turbulent Taylor vortices, while at $\mu = 0$ a featureless turbulent flow state develops inside the gap [15, 57, 82, 132]. In the following, this previous finding is confirmed within the present measurements and statistical profiles and velocity PDFs over cylindrical surfaces are compared to the literature for validation. Therefore, the azimuthally and time-averaged azimuthal velocity components

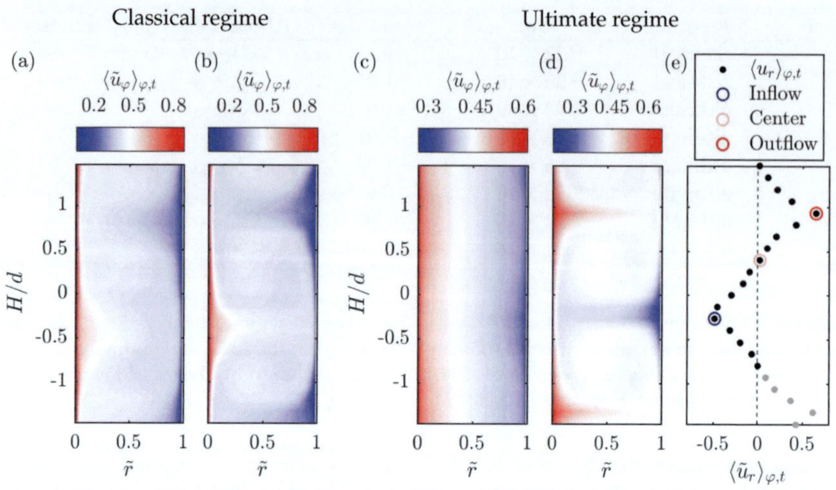

FIGURE 8.2: Flow states in terms of azimuthally and temporally (φ, t) averaged velocities. The contour plots depict the azimuthal velocity component for (a) C1$_{\mathrm{CR},0}$, (b) C2$_{\mathrm{CR,M}}$, (c) C6$_{\mathrm{UR},0}$ and (d) C7$_{\mathrm{UR,M}}$. (e) Axial profile of the radial velocity component for C7$_{\mathrm{UR,M}}$ at $\tilde{r} = 0.5$ with marked location of the vortex inflow, vortex center and vortex outflow. Gray dots represent data points, which are excluded for flow quantities calculated over the axial coordinate.

for the 23 investigated heights are shown in Figure 8.2. For both flows in the classical regime (see Figure 8.2(a,b)), large-scale rolls are visible in the mean field filling the whole gap. Apparently, the turbulent fluctuations depicted in Figure 8.4 are not strong enough at low shear Reynolds numbers to suppress these large-scale rolls. Further, the rolls are more pronounced at $\mu_{max} = -0.15$, indicating an increase in strength. In the ultimate regime at $Re_S = 2.15 \times 10^5$ and $\mu = 0$ (C6$_{\mathrm{UR},0}$), the turbulent Taylor rolls disappear and nearly no axial dependence of the azimuthal velocity is visible in the mean field, just as also found numerically (see the phase diagram, Figure 6 in Ostilla-Mónico et al. [86]). When the rotation rate is changed to μ_{max}, pronounced Taylor vortices are formed again, which are much more pronounced than in the classical regime. The contour plot reveals a mushroom-like structure with distinct in- and outflow regions. Compared to Figure 7.5, the large-scale rolls for $\eta = 0.5$ at $Re_S = 10^5$ looks more like the one presented here in the classical regime than in the ultimate regime for $\eta = 0.714$. The axial profile of the radial velocity component corresponding to Figure 8.2(d) is shown in Figure 8.2(e). This representation exemplifies the further analysis. The recorded 23 heights capture more than one vortex pair, which is why the grayly marked data points are excluded for flow quantities calculated over the axial coordinate. The axial length of evaluation therefore starts and ends at a vortex center. In between, the minimum of the azimuthally and temporally averaged axial profile of the radial velocity is used as the location of the vortex inflow and correspondingly, the location of the vortex outflow is obtained with its maximum. The data point, which is closest to a value of zero, is defined as the axial location of the vortex center.

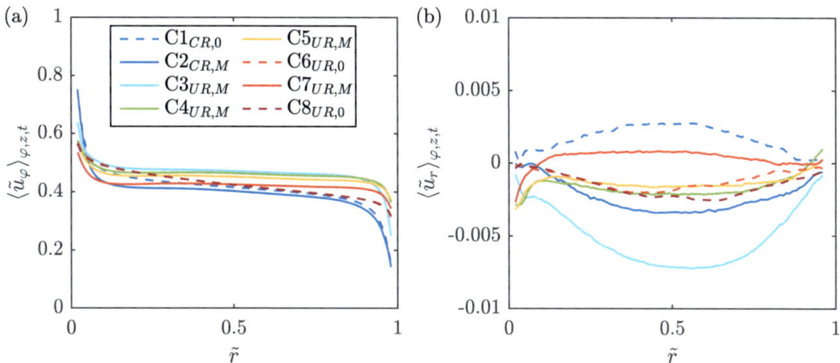

FIGURE 8.3: Radial profiles of the spatially and temporally (φ, z, t) averaged, normalized (a) azimuthal and (b) radial velocity component. The profiles are normalized according to Section 5.3.2.

In Figure 8.3, the normalized radial profiles of the azimuthal and radial velocity components are shown, averaged over space and time (φ, z, t). The slopes of the profiles in the bulk nearly vanish at $\mu = \mu_{max}$, while for pure inner cylinder rotation they show a small negative slope in good agreement with other studies [17, 45, 87]. With increasing shear Reynolds number, the difference of the data points close to the wall from the wall velocity becomes larger, which is due to the steepness of the velocity gradients. The averaged radial profiles of the radial velocity component are nearly zero all over the gap with absolute values smaller than 1% of the shear velocity u_S, as the radial velocity ideally has to vanish when averaged over one vortex pair. The deviation results from the restricted axial resolution of the individual heights.

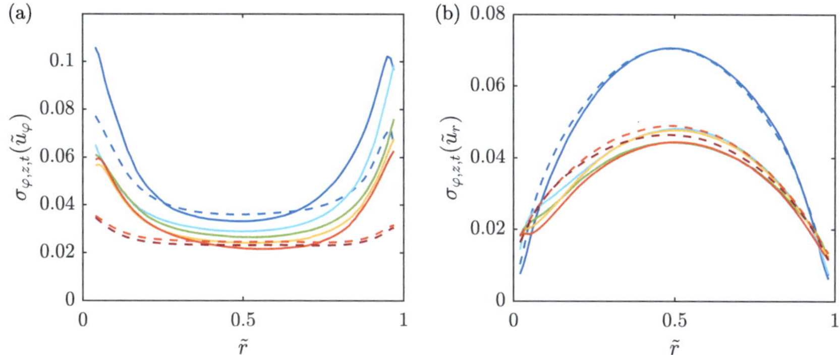

FIGURE 8.4: Radial profiles of the standard deviation of the (a) azimuthal and (b) radial velocity component calculated over space and time (φ, z, t). The profiles are normalized by the shear velocity u_S. Colors are chosen according to table 8.1.

Next, the radial profiles of the standard deviation of the azimuthal and radial velocity component, calculated over space and time (φ, z, t), are plotted in Figure 8.4. In terms of the azimuthal velocity component, the radial profiles of the standard deviation show a nearly constant low value in the center of the gap and increasing values close to the wall. This increase is more pronounced at μ_{max}; while for the lowest shear Reynolds number, a peak close to the outer cylinder wall is visible. In the case of the radial velocity component, the standard deviation depicts a maximum in the center of the gap and decreases towards the cylinder walls. The maximum of $\sigma_{\varphi,z,t}(\tilde{u}_r)$ for both flow cases in the classical regime is noticeably higher than the one in the ultimate regime. The overall shape of the profiles agrees well with the ones for pure inner cylinder rotation of Ezeta *et al.* [37], measured with PIV in the same facility at mid-height. Further, it can be stressed that the averaged flow structure and profiles present the same characteristics as the one found for $\eta = 0.5$ in Section 7.2.

8.2.1   Probability density functions (PDFs) of velocity components

To provide further validation of the here discussed measurements and a basis for the investigation of the local PDFs of the angular momentum transport (see Section 8.4), firstly the PDFs of the azimuthal and radial velocity components are analyzed.

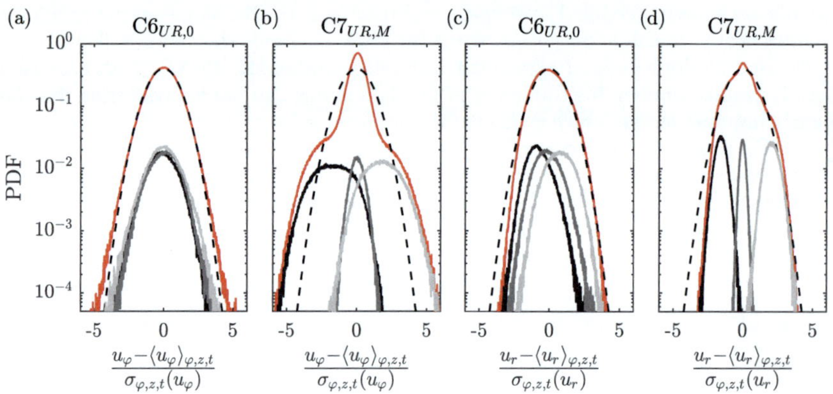

$$\frac{u_\varphi - \langle u_\varphi \rangle_{\varphi,z,t}}{\sigma_{\varphi,z,t}(u_\varphi)} \qquad \frac{u_\varphi - \langle u_\varphi \rangle_{\varphi,z,t}}{\sigma_{\varphi,z,t}(u_\varphi)} \qquad \frac{u_r - \langle u_r \rangle_{\varphi,z,t}}{\sigma_{\varphi,z,t}(u_r)} \qquad \frac{u_r - \langle u_r \rangle_{\varphi,z,t}}{\sigma_{\varphi,z,t}(u_r)}$$

FIGURE 8.5: PDFs of the (a,b) azimuthal and (c,d) radial velocity component calculated over space and time (φ, z, t) at $\tilde{r} = 0.5$ for $Re_S = 2.1 \times 10^5$ and $\mu = 0$ as well as μ_{max}. Red dashed lines correspond to $\mu = 0$ (C6$_{UR,0}$) and red solid lines to $\mu = -0.36$ (C7$_{UR,M}$). Dark gray, gray and bright gray lines indicate the local PDFs for the vortex inflow, vortex center and vortex outflow, respectively. The black dashed line represents a Gaussian distribution with zero mean and unit variance.

In Figure 8.5, the PDFs of the azimuthal and radial velocity components are shown at $\tilde{r} = 0.5$ and $Re_S = 2.1 \times 10^5$ for $\mu = 0$ (C6$_{UR,0}$) and $\mu = -0.36$ (C7$_{UR,M}$) as representatives. In addition, the underlying PDFs at the locations of the vortex inflow, vortex center and vortex outflow are included. The local PDFs at the specific vortex locations are close to Gaussian distributions, as already shown by Huisman *et al.* [58] for the azimuthal velocity component, measured via LDV at mid-height and $\tilde{r} = 0.5$. When all data at different heights are included in the PDFs, they still follow a Gaussian shape at $\mu = 0$ for both velocity components. At μ_{max} however, the PDF of the azimuthal velocity depicts a

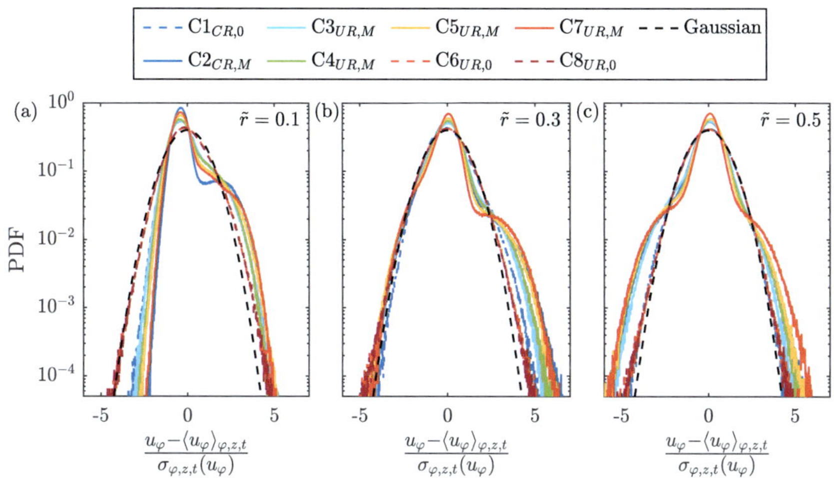

FIGURE 8.6: PDFs of the azimuthal velocity component calculated over space and time (φ, z, t) for the radial locations (a) $\tilde{r} = 0.1$, (b) $\tilde{r} = 0.3$ and (c) $\tilde{r} = 0.5$

cusp-like form centered at the origin with approximately exponential tails in accordance with the numerical simulations at a lower Reynolds number of $Re_S = 2 \times 10^4$ by Brauckmann *et al.* [17]. There, and in the study of Emran and Schumacher [35], a nearly identical shape was reported for the temperature PDF in RB flow, which reveals yet another clear evidence of the analogies between both systems, even in the small-scale statistics. In RB flow, the specific PDF shape of the temperature is induced by a combination of bursting plumes and large-scale rolls [20, 98, 143]. By using the TC-RB flow analogy, here a clear fingerprint of plumes transporting angular momentum is identified in TC flow. In addition, the vortex-dominated TC flow in the region of the torque maximum in the fully turbulent regime shows a similar behavior to the flow organization in RB flow. In the case of the radial velocity at μ_{max}, the PDF also deviates from the ideal Gaussian shape, which can be attributed again to the intermittent bursting plumes.

As the PDFs of u_r for flow cases at pure inner cylinder rotation depict a nearly Gaussian shape, which is also valid for different radial locations in the bulk, the radial velocity component is omitted for the following analysis within this section. In Figure 8.6, the PDF of the azimuthal velocity component is calculated for different radial locations. When the point of evaluation approaches from the gap center to the direction of the inner cylinder wall for μ_{max}, the right-hand tail of the initial cusp-like PDF becomes more pronounced and its width increases with the shear Reynolds number. This change is caused by the dominance of the vortex outflow in the inner gap region, where the mean azimuthal velocity exhibits a large positive value. In addition, very close to the inner wall at $\tilde{r} = 0.1$, both tails become increasingly exponential. This behavior is even more pronounced at higher Re_S, which is another sign of the ejection of coherent plumes from the cylinder wall. Next to these local changes, the global asymmetry of the PDFs seems to increase with decreasing distance to the wall.

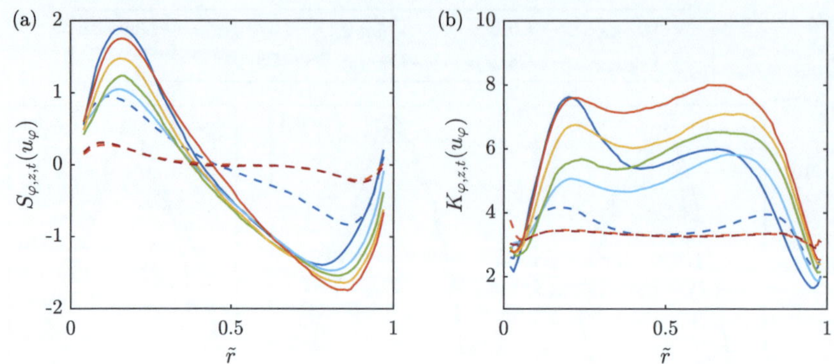

FIGURE 8.7: Radial profiles of the (a) skewness S and (b) kurtosis K of the azimuthal velocity component calculated over space and time (φ, z, t). Colors are chosen according to table 8.1.

To account for such global properties of the PDFs, the radial profiles of skewness S and kurtosis K of the azimuthal velocity component are shown in Figure 8.7 for different Reynolds numbers. When the outer cylinder is at rest and the ultimate regime is reached, the skewness is close to zero and the kurtosis close to three, confirming the nearly Gaussian shape. The small radius dependent deviations in skewness for large Re_S may result from remnants of turbulent Taylor vortices [57, 66, 128]. At μ_{max} the skewness increases from the inner cylinder wall to a maximum around $\tilde{r} = 0.16$, then decreases to negative values with a minimum around $\tilde{r} = 0.86$ for $Re_S = 2.14 \times 10^5$ (C7$_{\text{UR,M}}$), and then increases again in the direction of the outer cylinder. Furthermore, the absolute skewness increases with the shear Reynolds number. This behavior is similar to that of the temperature fields in RB flows reported by Emran and Schumacher [35]. The kurtosis profiles at μ_{max} depict two maxima, one in the inner gap region at $\tilde{r} = 0.22$ and one in the outer gap region at $\tilde{r} = 0.68$ for the highest Re_S. This reflects that the non-Gaussianity of the PDFs is most pronounced in the regions dominated by the vortex in- and outflows due to coherent plumes. In summary, the global and local velocity field statistics of the present measurements agree very well with those of the mentioned literature and exceed the state-of-the-art especially for μ_{max} to higher forcings.

8.3   Local and global angular momentum transport

As the angular momentum transport, expressed in terms of the pseudo-Nusselt number Nu_ω, is strongly influenced by the local and global flow organization inside the gap, it is an appropriate parameter to statistically investigate the existence of flow structures. Therefore, within this section, first the global angular momentum transport is analyzed to compare its amount with the literature, and second, the axial dependent radial profiles of Nu_ω are analyzed to reveal the most relevant axial locations for the transport. The subsequent results are fundamental for the small-scale statistical analysis of the local momentum transport in the next section and enable new insights into the vortex-dominated momentum transport. The Nusselt number is composed of a convective and a viscous part [33],

$$\mathrm{Nu}_\omega = \mathrm{Nu}_\omega^c(r) + \mathrm{Nu}_\omega^v(r) = \frac{r^2}{J_\omega^{lam}} \left\langle u_\varphi u_r \right\rangle_{\varphi,z,t} - \frac{\nu r^3}{J_\omega^{lam}} \partial_r \left\langle \frac{u_\varphi}{r} \right\rangle_{\varphi,z,t}. \tag{8.1}$$

While the viscous term Nu_ω^v dominates in the boundary layers, the convective term Nu_ω^c dominates in the bulk. Since the focus of the present investigation is mainly in the bulk region, the viscous part is neglected. Furthermore, as is shown in Figure 8.3(b), the radial velocity component nearly vanishes when averaged over one vortex pair: $\langle u_r \rangle_{\varphi,z,t} \approx 0$. Therefore, $r^2 \langle u_r u_\varphi \rangle_{\varphi,z,t} \approx r^2 \langle u_r''' u_\varphi''' \rangle_{\varphi,z,t}$ is valid, which means that only the fluctuations of the azimuthal velocity component around its mean profile contribute to the net momentum flux through these cylindrical surfaces [17]. Accordingly, the net convective flux in the bulk flow can be calculated as

$$\mathrm{Nu}_\omega^{c,net} = \frac{r^2}{J_\omega^{lam}} \left\langle u_\varphi''' u_r''' \right\rangle_{\varphi,z,t}, \quad \text{with} \tag{8.2}$$

$$u_r'''(r,\varphi,z,t) = u_r(r,\varphi,z,t) - \langle u_r(r,\varphi,z,t) \rangle_{\varphi,z,t},$$
$$u_\varphi'''(r,\varphi,z,t) = u_\varphi(r,\varphi,z,t) - \langle u_\varphi(r,\varphi,z,t) \rangle_{\varphi,z,t}.$$

To avoid confusion, the Nusselt number $\mathrm{Nu}_\omega = \mathrm{Nu}_\omega^c + \mathrm{Nu}_\omega^v$ shown in equation (8.1) is called the *total* Nusselt number in the following. In Figure 8.8(a), the radial profiles of Nu_ω, indicated by lines, and the net convective momentum flux $\mathrm{Nu}_\omega^{c,net}$, indicated by triangles (μ_{max}) and circles ($\mu = 0$), are shown. The analysis is restricted to $\tilde{r} \in [0.1, 0.9]$.

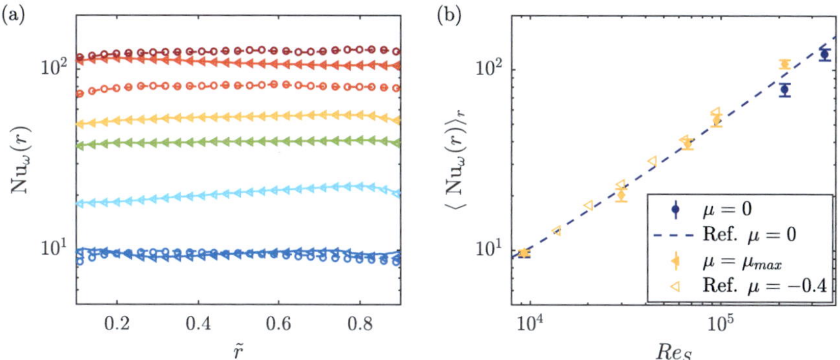

FIGURE 8.8: (a) Radial profiles of the total Nusselt number (lines) and of the net convective Nusselt number (triangles for μ_{max}, circles for $\mu = 0$). The evaluation is restricted to the interval $0.1 \leq \tilde{r} \leq 0.9$ to exclude the boundary layers and focus on the bulk flow. Colors are chosen according to table 8.1. (b) Radially averaged total Nusselt number as function of Re_S. Errorbars represent the standard deviation of the total Nusselt number along the radial coordinate. Data are compared to torque measurements of Lewis and Swinney [68] (Ref. $\mu = 0$, $\eta = 0.724$) and DNSs of Ostilla-Mónico *et al.* [86] (Ref. $\mu = -0.4$, $\eta = 0.714$)

A slight dependence of the Nu_ω-profiles on the radial coordinate is observable, which differs partially for the different flow states. The deviation from the predicted conservation of the momentum transport along the radial direction according to Eckhardt *et al.*

[33] is probably due to the finite axial length of the used experimental setup in contrast to their theory, which is based on cylinders of infinite length. However, the effect of the cylinder length is reduced in the present setup by cutting the axial length of evaluation to the size of one vortex pair. Moreover, also the vortex aspect ratio influences the value of Nu_ω as a function of Re_S [57, 74], which takes values for the current study in the range of $2.13d - 2.67d$. Considering these aspects, the shape of the Nu_ω-profiles is satisfactory. The net convective momentum flux dominates the total Nusselt number and is valid to evaluate the momentum transport in the bulk region [60]. Furthermore, in Figure 8.8(b), the radially averaged total Nusselt number is shown as a function of Re_S, which is also compared to both torque measurements of Lewis and Swinney [68] ($\mu = 0, \eta = 0.724$) and DNSs of Ostilla-Mónico *et al.* [86] ($\mu = -0.4, \eta = 0.714$). A very good agreement is found with these data, which enables a more detailed analysis of the momentum transport.

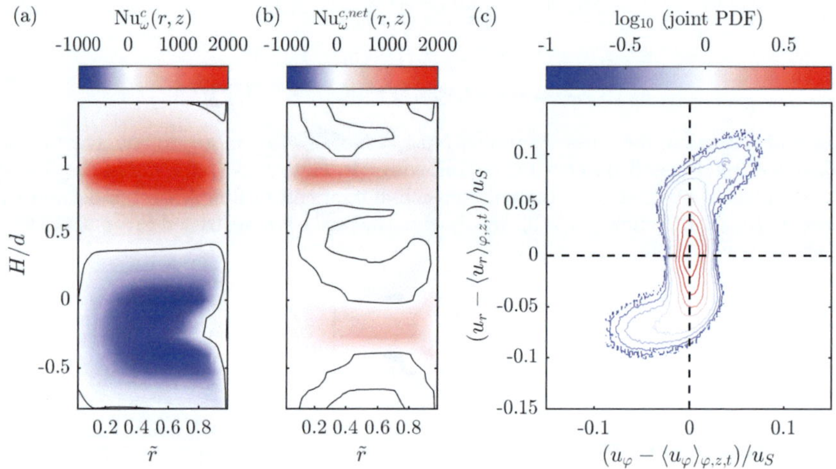

FIGURE 8.9: Contour plot of the dimensionless (a) convective and (b) net convective local angular momentum transport in a meridonal plane for $Re_S = 2.14 \times 10^5$ and $\mu = -0.36$ (C7$_{UR,M}$). Black lines indicate the zero line of the depicted quantities. The color codes are given in the legends. (c) Joint PDF of the radial and azimuthal velocity fluctuations for the same flow state at $\tilde{r} = 0.5$. The colors give the probability (in log-scale), see legend.

In order to get a deeper insight into the relation between the convective and net convective Nusselt number, both local quantities are plotted in a meridional plane ($r - z$ plane) for $Re_S = 2.14 \times 10^4$ and $\mu = -0.36$ (C7$_{UR,M}$) in Figure 8.9(a,b). The local Nusselt number can be much larger than its average value as already noticed by Huisman *et al.* [60] and Ostilla-Mónico *et al.* [83]. In the case of the convective Nusselt number Nu_ω^c (Figure 8.9(a)), the momentum transport in the area of the vortex outflow is positive and strongly concentrated in a small axial region, while in the area of the vortex inflow, the transport is negative and less focused. The magnitude of positive momentum flux is twice as large as the negative one, which yields on average a positive net transport. Alternatively, when the net transport is plotted (Figure 8.9(b)), a much clearer picture of the transport process is revealed. While in the in- and outflow region the net transport is positive, only in the sheared regions in between negative net transport can be detected. In order to explain this difference, the joint PDF of the radial and azimuthal velocity fluctuations is shown at $\tilde{r} = 0.5$ in Figure 8.9(c). In the inflow region, where the fluid is transported strongly in the negative radial direction, the azimuthal velocity depicts a high probability for negative fluctuation values. As a consequence, the net transport has to be positive. In the outflow region, both velocities are mainly positive, resulting also in a positive correlation. However, when the mean azimuthal velocity component is not subtracted, the joint PDF is shifted to the right, leading to negative correlations in the inflow region (not shown). In summary, the difference between Nu_ω^c and $\mathrm{Nu}_\omega^{c,net}$ is the transport of the mean azimuthal velocity by the turbulent Taylor vortices, which vanishes when averaged over cylindrical surfaces. By neglecting this fraction, a much clearer picture of the transport process is revealed. In addition, the representation of the Nusselt number in the meridional plane demonstrates the importance of the in- and outflow regions for the net convective transport.

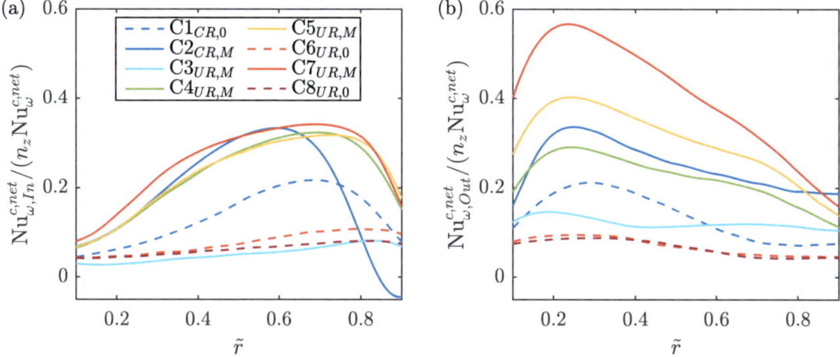

FIGURE 8.10: Radial profiles of the contribution of the net convective momentum transport of the (a) inflow and (b) outflow to the total transport. n_z represents the number of heights included into the average over cylindrical surfaces.

The contribution of the vortex in- and outflow as a function of the radial location and Re_S is depicted in Figure 8.10. As a global feature, the contribution of the vortex inflow to the total net convective momentum transport is especially pronounced in the outer gap region ($\tilde{r} > 0.5$) while the opposite is true for the outflow. For the two flow states in the classical regime, where the values of the total Nusselt numbers are comparable (see Figure 8.8), the contributions of the in- and outflow are much more dominant at μ_{max}. This reflects the strong correlation of the enhanced momentum flux at μ_{max} and the turbulent Taylor vortices. Further, in the ultimate regime at $\mu = 0$, again the effect of remnants of these large vortices becomes visible in the slight dependence of $\mathrm{Nu}_\omega^{c,net}$ on $\tilde{r}$. When the ultimate regime is reached at $Re_S = 2.98 \times 10^4$ at μ_{max} (C3$_{UR,M}$), the contributions of the vortex in- and outflow are small and become significant again at $Re_S = 6.68 \times 10^4$ (C4$_{UR,M}$). For even higher shear Reynolds numbers, the contribution of the inflow stays nearly constant with a fraction slightly above 30% in the outer gap region; while the outflow contribution continuously increases up to approximately 60% in the inner gap region. This value is strikingly high and demonstrates that at very high Reynolds numbers, the net convective transport shrinks to a very small axial region, where most of the momentum transport takes place. It should be noted that further studies would be desirable to confirm this finding.

8.4   PDFs of the net convective angular momentum flux

With the knowledge of the total net convective transport and the relative contributions of the in- and outflow, the PDFs of $\mathrm{Nu}_\omega^{c,net}$ are now analyzed to identify statistical footprints of small-scale plume structures. To properly normalize these PDFs and provide an idealized shape for comparison, at first the Gaussian distribution is introduced. According to Figure 8.5, the PDFs of the radial and azimuthal velocity components are nearly Gaussian for pure inner cylinder rotation. In that case, the PDFs of their product can be described according to Chu *et al.* [22] and Thoroddsen and van Atta [122] based on a Gaussian distribution. The PDF of two jointly Gaussian random variables x and y is given by

$$PDF(x,y) = \frac{1}{\pi \sigma_x \sigma_y \sqrt{1 - \rho_P^2}} \exp\left[\frac{1}{2(1 - \rho_P^2)} \left(\frac{x^2}{\sigma_x^2} - \frac{2xy\rho_P}{\sigma_x \sigma_y} + \frac{y^2}{\sigma_y^2} \right) \right], \qquad (8.3)$$

with the individual standard deviations σ_x and σ_y and the associated correlation coefficient $\rho_P = \langle xy \rangle / (\sigma_x \sigma_y)$. Equation (8.3) represents an inclined elliptic shape for the joint PDF in contrast to the one of Figure 8.9(c). Furthermore, the PDF of the product $z = xy$ is [22, 122]

$$PDF(z) = \frac{1}{\pi \sigma_x \sigma_y \sqrt{1 - \rho_P^2}} \exp\left(\frac{\rho_P z}{(1 - \rho_P^2)\, \sigma_x \sigma_y} \right) K_0 \left(\frac{|z|}{(1 - \rho_P^2)\, \sigma_x \sigma_y} \right), \qquad (8.4)$$

with K_0 the modified Bessel function of the second kind. The prefactor of the exponential function in equation (8.4) is used to normalize the different PDFs of the angular momentum flux, i.e to standardize the width of their tails. It is worth mentioning that the correlation coefficient ρ_P cannot be normalized in a proper way, which leads to different slopes of the exponential tails of the prediction depending on ρ_P. Therefore, the prediction according to equation (8.4) is used in the present calculation for the case C8$_{\mathrm{UR,0}}$. In Figure 8.11, ρ_P is depicted for all flow cases as a reference. The correlation between the azimuthal and radial velocity component in the ultimate regime decreases with the radial coordinate for μ_{max}, while a local minimum at $\tilde{r} \approx 0.6$ can be seen in case of $\mu = 0$. It can be observed that ρ_P in a flow with strong Taylor rolls, is much larger than in a flow without them.

In Figure 8.12, the PDFs of $\mathrm{Nu}_\omega^{c,net}$ are shown at $\tilde{r} = 0.5$, which are normalized with the factor $\sigma_u \sigma_{u_\theta} \sqrt{1 - \rho_P^2}$ in order to compare their shapes. For pure inner cylinder rotation, the PDFs of the net convective momentum transport agree very well with the proposed jointly Gaussian prediction. However, at μ_{max} the PDFs become highly skewed due to a change of shape in the positive tails. This deviation results in an increasing but still small number of positive extreme and rare events of momentum flux, which becomes more pronounced with increasing Re_S. According to Figure 8.12(c), these rare and extreme events can be almost completely attributed to the vortex in- and especially the vortex outflow

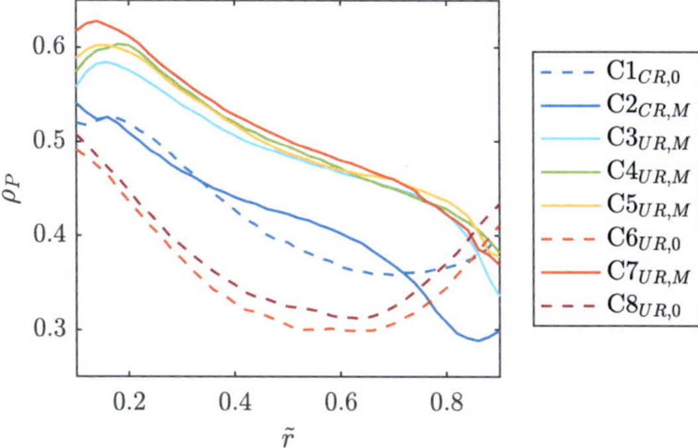

FIGURE 8.11: Correlation coefficient ρ_P of the radial and azimuthal velocity components calculated over space and time (φ, z, t) as a function of the radial coordinate. Note that the correlation coefficient related to Figure 8.9(c) for $Re_S = 2.14 \times 10^5$ and $\mu = -0.36$ (C7$_{\mathrm{UR,M}}$) becomes $\rho_P = 0.502$ at $\tilde{r} = 0.5$.

due to changes in the azimuthal velocity component (see also Figure 8.5). Here, a strong correlation between the radial and azimuthal velocity component exists due to the coherence of the plumes (see also Figure 8.9(c)). In addition, all PDFs have in common that zero is the value with the highest probability instead of the average value of the Nusselt number.

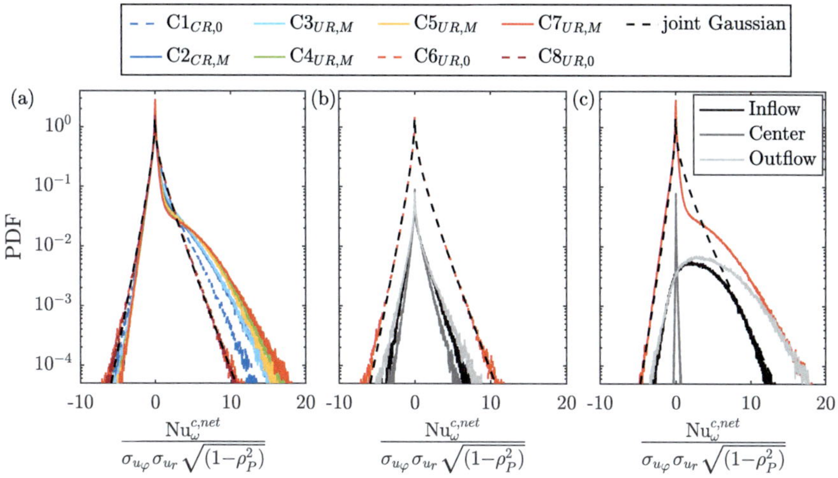

FIGURE 8.12: PDFs of the local net convective angular momentum transport calculated over space and time (φ, z, t) for $\tilde{r} = 0.5$. (a) PDFs of all investigated flow states. (b) PDF for $Re_S = 2.15 \times 10^5$ and $\mu = 0$ (C6$_{\mathrm{UR,0}}$) with the corresponding PDFs at the axial height of the vortex inflow, vortex center and vortex outflow. (c) Same plot as in (b) for $Re_S = 2.14 \times 10^5$ and $\mu = -0.36$ (C7$_{\mathrm{UR,M}}$). The dashed black line indicates the prediction according to equation (8.4).

The overall shape of the PDFs at μ_{max} is in good agreement with the findings of Brauck-mann *et al.* [16], whose analysis was however restricted to low Reynolds numbers and PDFs along cylinder surfaces ($Re_S = 2 \times 10^4$). Furthermore the PDFs of the current study are comparable to the corresponding PDFs of the heat flux in the RB flow. Shang *et al.* [109] reported similar PDF shapes in the case of RB convection. They argued that such large rare events are footprints of heat flux fluctuations induced by thermal plumes. This feature is obviously shared with TC flows and supports the origin of large-scale turbulent Taylor vortices as the result of small-scale unmixed plumes. Even more, slight counter-rotating cylinders at μ_{max} instead of pure inner cylinder rotation seems to be the right kinematic boundary condition of TC flows for comparisons with the RB flow concerning Nu_ω-PDFs.

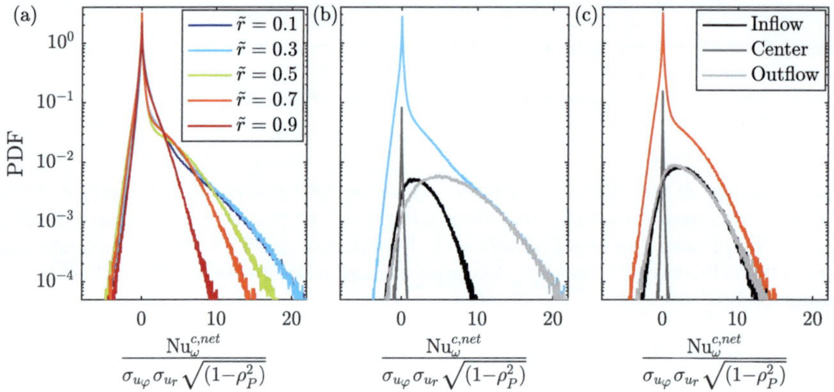

FIGURE 8.13: (a) PDFs of the local net convective angular momentum transport calculated over space and time (φ, z, t) for $Re_S = 2.14 \times 10^5$, $\mu = -0.36$ (C7$_{\mathrm{UR,M}}$) and different radial positions. (b,c) Same PDFs at $\tilde{r} = 0.3$ and $\tilde{r} = 0.7$, respectively, with the corresponding PDFs related to the vortex inflow, vortex center, and vortex outflow.

In Figure 8.13, the PDFs of $\mathrm{Nu}_\omega^{c,net}$ are shown for various different radii r and the case of $Re_S = 2.14 \times 10^5$ at $\mu = -0.36$ (C7$_{\mathrm{UR,M}}$). The negative tails of the PDFs are largely identical for all investigated radial positions, while the right tails strongly depend on $\tilde{r}$. From the inner to the outer cylinder wall, the width of the positive tail and therefore the asymmetry decreases coinciding with degeneration of the exponential tails in the u_φ-PDFs in Figure 8.6. As was shown in Figure 8.10(b), the maximum of the outflow contribution to the overall momentum transport is close to $\tilde{r} = 0.3$. At this location, the right tail of the PDF reflects extreme positive events due to the emission of plumes and is nearly covered by the PDF of the outflow (see Figure 8.13(b)). However, in the outer gap region at $\tilde{r} = 0.7$, the asymmetric right tail of the PDF is comparably formed by both the in- and outflow region. The findings of this section demonstrate that the comparative evaluation of axially-dependent global and local PDFs of $\mathrm{Nu}_\omega^{c,net}$, which has not yet been done in TC flow before, is crucial for deeper insights into the statistics of the momentum transport.

8.5 Azimuthal energy co-spectra and correlations

8.5.1 Azimuthal energy co-spectra

As shown in the previous section, the PDFs of the net convective Nusselt numbers suggest the presence of small-scale plumes concentrated especially in the in- and outflow regions of the turbulent Taylor vortices, which dominate the transport. Hence, the spatial energy co-spectra are analyzed to detect the presence and length scale of small-scale structures in the gap. The algorithm is according to Press *et al.* [96]. Velocity fluctuations can be assumed at a constant radial r_c and axial z_c position in the homogeneous φ-direction with a total number of azimuthal points of $n_\varphi = 0, 1, ..., N - 1$ equidistantly spaced by the arc length interval $\Delta s = \Delta \varphi r$, i.e.

$$u'_r (r_c, \varphi, z_c, t) = u_r (r_c, \varphi, z_c, t) - \langle u_r (r_c, \varphi, z_c, t) \rangle_t, \tag{8.5}$$

$$u'_\varphi (r_c, \varphi, z_c, t) = u_\varphi (r_c, \varphi, z_c, t) - \langle u_\varphi (r_c, \varphi, z_c, t) \rangle_t. \tag{8.6}$$

The discrete spatial Fourier transform $\hat{U}_{r,\varphi}$ of both fluctuation components $u'_{r,\varphi}$ is given by

$$\hat{U}_{r,\varphi} (n_\varphi) = \sum_{k=0}^{N-1} u'_{r,\varphi} (k) \exp \left(-\frac{2\pi i k n_\varphi}{N} \right), \tag{8.7}$$

where for simplicity the dependences on r_c, z_c and t. have not been written out. Thus, the spatial energy co-spectrum $E_{r\varphi}$ can be calculated as

$$E_{r\varphi} \left(k_\varphi^{n_\varphi} \right) = \begin{cases} \frac{1}{N^2} |\hat{U}_r^{n_\varphi} \cdot \hat{U}_\varphi^{n_\varphi}| & \text{for } n_\varphi = [0, \frac{N}{2}], \\ \frac{1}{N^2} \left(|\hat{U}_r^{n_\varphi} \cdot \hat{U}_\varphi^{n_\varphi}| + |\hat{U}_r^{N-n_\varphi} \cdot \hat{U}_\varphi^{N-n_\varphi}| \right) & \text{for } n_\varphi = [1, ..., \frac{N}{2} - 1], \end{cases} \tag{8.8}$$

with the wavenumber vector $k_\varphi^{n_\varphi} = (\Delta s)^{-1} n_\varphi / N$. The co-spectra are determined for each time step t and afterwards ensemble averaged over 1500 snapshots for every case. To enable a comparison of the co-spectra for different Re_S and radial positions, all co-spectra are normalized with the area A_E below the corresponding graph, based on the trapezoidal integration method: $A_E \approx (2\Delta s)^{-1} \sum_{n_\varphi=0}^{N/2-1} \left[E_{r\varphi}(k_\varphi^{n_\varphi}) + E_{r\varphi}(k_\varphi^{n_\varphi+1}) \right]$.

First of all, the temporally and axially averaged azimuthal energy co-spectra are shown at $\tilde{r} = 0.5$ in Figure 8.14(a) to illustrate their scaling and compare it with other studies. The kinetic energy co-spectra show that most of the energy lies within the large scales corresponding to small wavenumbers, and depict a noticeable drop for $k_\varphi d \approx 20$. It is worth mentioning that it cannot be seen any peak in the large-scale regime because of the limited range of the azimuthal coordinate in the experiments. Moreover, the energy content of axisymmetric Taylor rolls cannot be resolved, as they correspond to an azimuthal wavenumber of $k_\varphi = 0$. In the case of Re_S being in the classical regime, the spectra show, compared to the other flow states, a stronger decrease of the spectral energy at mid scales and a kink in the region of $10 \leq k_\varphi d \leq 20$. When Re_S is increased into the ultimate

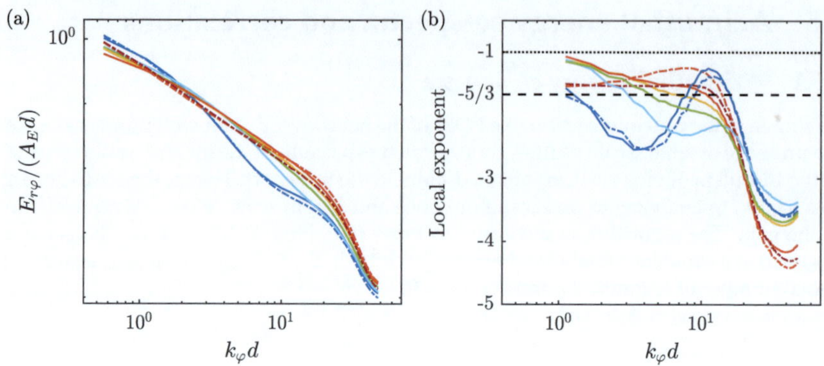

FIGURE 8.14: (a) Temporally and azimuthally averaged azimuthal kinetic energy co-spectra evaluated at $\tilde{r} = 0.5$. Spectra are normalized to cover an area of 1 under their respective curves and by the gap width d. (b) Local scaling exponent γ of the co-spectra for $E_{r\varphi} \sim k_\varphi^\gamma$ calculated with a bin size of of $\log_{10}(k_\varphi) = 0.5$. Colors are chosen according to table 8.1.

regime (with μ fixed at $\mu = \mu_{max}$), this kink continuously diminishes and energy is redistributed from the large to the small scales. Based on the power-law ansatz $E_{r\varphi} \sim k_\varphi^\gamma$, the local scaling exponent γ corresponding to the co-spectra is shown in Figure 8.14(b); γ is calculated with a bin size of $\log_{10}(k_\varphi) = 0.5$. It is worth mentioning that γ is sensitive to the data processing and the accompanying error propagation: γ is given here for this choice of bin size and to show the trend of γ with d. It can be observed that the spectra neither show -1 nor $-5/3$ scaling, which is consistent with the results of Huisman *et al.* [58], Lewis and Swinney [68] and Ostilla-Mónico *et al.* [88]. In the ultimate regime (with μ fixed at $\mu = \mu_{max}$), the exponent decreases with increasing k_φ before it strongly drops down in the viscous regime beyond $k_\varphi d \approx 20$. This decrease becomes smaller with increasing Re_S. However, in the case of the highest investigated shear Reynolds number at $Re_S = 3.51 \times 10^5$ and $\mu = 0$ (C8$_{UR,0}$), the local exponent is nearly constant for $k_\varphi d \leq 10$ with a value of $\gamma \approx -1.52$, which is slightly above $-5/3$.

Next, the focus is set to the local pre-multiplied energy co-spectra at the axial height of the vortex inflow, vortex center, and vortex outflow, respectively, where the PDF analysis of $\mathrm{Nu}_\omega^{c,net}$ suggests the occurrence of small-scale intermittent plumes. *Pre-multiplied* means that the co-spectra are multiplied with the wavenumber vector k_φ, such that the area under its graph corresponds to the kinetic energy [112]. In Figure 8.15, the premultiplied energy co-spectra are shown in the classical regime for $Re_S = 9.32 \times 10^3$ and μ_{max} (C2$_{UR,M}$) at the three vortex positions. Firstly, the peak is addressed that corresponds to the large scales. For the vortex inflow in Figure 8.15(a), the large-scale peak is located around $k_\varphi d \approx 1.1$ within the bulk for $0.2 \leq \tilde{r} \leq 0.8$, while close to the outer cylinder at $\tilde{r} = 0.9$, the peak lies outside of the resolvable scales. Close to the IC wall however, the peak shifts to $k_\varphi d \approx 2.2$ at $\tilde{r} = 0.1$. In Figure 8.15(c), at the location of the vortex outflow, the opposite behavior is observed. Here, the large-scale peak shifts to smaller wavenumbers, when the radial position increases from the IC to the OC. At the vortex center in Figure 8.15(b), the large-scale peak is shifted from smaller wavenumbers in the center of the gap ($\tilde{r} = 0.5$) to larger ones at both cylinder walls, in an almost symmetric manner.

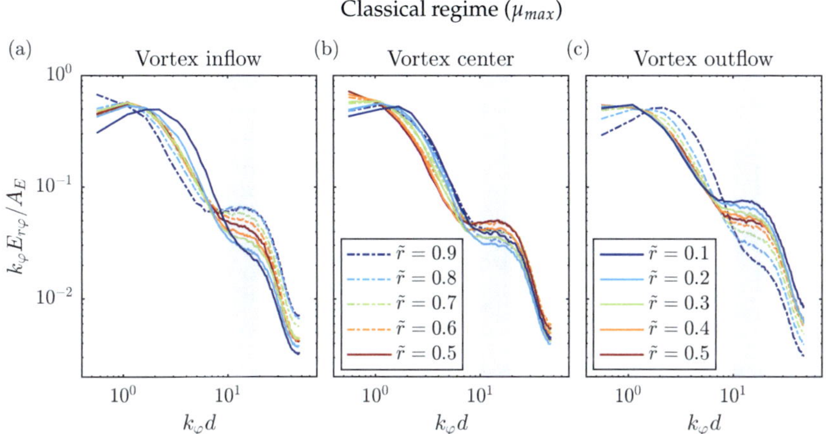

FIGURE 8.15: Temporally averaged pre-multiplied azimuthal kinetic energy co-spectra in the classical regime for $Re_S = 9.30 \times 10^3$ and $\mu = -0.15$ (C2$_{\text{CR,M}}$) at different radial positions at the axial height of (a) the vortex inflow, (b) the vortex center and (c) the vortex outflow. The region of $k_\varphi d \in [10, 20]$ with enhanced plume emission is marked in light blue.

This suggests that the formation of this large-scale peak is connected to the mean radial velocity field, which is in itself caused by the large-scale turbulent Taylor rolls. When fluid impacts on the cylinder walls due to these rolls, structures of the size $k_\varphi d \approx 2.2$ are formed. Since the axisymmetric flow has a wavenumber of $k_\varphi = 0$, the existence of a large-scale peak may be an indication of modified turbulent Taylor vortices. This point will be discussed in more detail in Section 8.6.

With respect to the small-scale peak, it can be observed that it is located at wavenumbers around $k_\varphi d \in [10, 20]$ for all three depicted heights, independent of the radial coordinate. However, its amplitude strongly varies with $\tilde{r}$ and the height z. In the region of the vortex inflow shown in Figure 8.15(a), the small-scale peak is most pronounced near the OC and its amplitude decreases monotonically with decreasing $\tilde{r}$. On the contrary, at the height of the vortex outflow in Figure 8.15(c), the small-scale peak amplitude is largest close to the IC and decreases in amplitude towards the OC. In the region of the vortex center in Figure 8.15(c), the highest amplitude of the small-scale peak is found in the center of the gap. This is due to the emission of coherent plumes from the cylinder walls, which give rise to the formation of Taylor rolls, as was already mentioned before. Thus, this peak should indeed be most pronounced in the ejecting regions, i.e. in the outflow region at the IC and in the inflow region at the OC, respectively. At the height of the vortex center —where no predominant flow direction concerning the radial velocity component is present— the behavior is different: plumes rise from both cylinder walls and travel towards the bulk flow, which results in the highest peak amplitude at $\tilde{r} = 0.5$. Considering the fact, that the contribution of the vortex center to extreme and strong events of momentum flux is almost negligible (see Figures 8.12, 8.13), these detected plumes seem to compensate their total radial momentum transport.

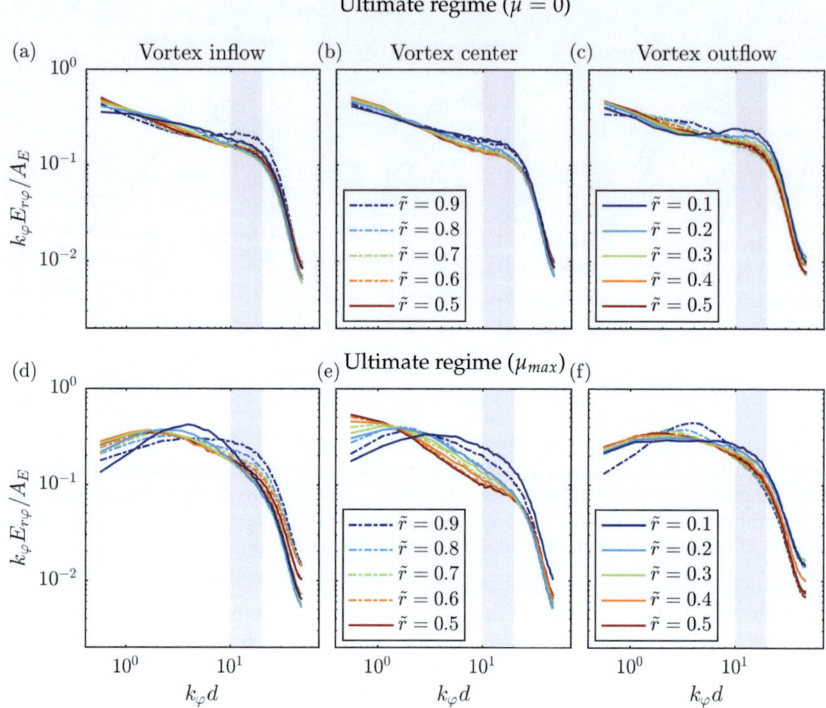

FIGURE 8.16: Upper row: Temporally averaged pre-multiplied azimuthal kinetic energy co-spectra for $Re_S = 2.15 \times 10^5$ and $\mu = 0$ (C6$_{\text{UR},0}$) at different radial positions at the axial height of (a) the vortex inflow, (b) the vortex center and (c) the vortex outflow. Region of $k_\varphi d \in [10, 20]$ is marked in light blue. Lower row: Same spectra for $Re_S = 2.14 \times 10^5$ and $\mu = -0.36$ (C7$_{\text{UR},\text{M}}$) at the axial height of (d) the vortex inflow, (e) the vortex center and (f) the vortex outflow.

The pre-multiplied energy co-spectra in the ultimate regime at $Re_S = 2.1 \times 10^5$ for the three vortex locations are depicted in Figure 8.16 for $\mu = 0$ (C6$_{\text{UR},0}$, a-c) and μ_{max} (C7$_{\text{UR},\text{M}}$, d-f). For $\mu = 0$, no large-scale peak can be seen in any of the co-spectra at the investigated heights. This is consistent with the finding shown in Figure 8.2(c), where it is observed that the Taylor rolls have faded away. Accordingly, the shape of the co-spectra is less dependent on both the axial coordinate and on $\tilde{r}$. However, the co-spectra show a prominent change in the slope around $k_\varphi d \approx 20$. In Figure 8.16(a) (vortex inflow), a small-scale peak is formed around $k_\varphi d \in [10, 20]$ close to the OC at $\tilde{r} = 0.9$. This is comparable to the previous case in the classical regime. Also in the region of the vortex outflow (Figure 8.16(c)), a peak is observed within the same range of scales at $\tilde{r} = 0.1$. For the vortex center however (Figure 8.16(b)), no peak is visible.

Also when μ is changed to μ_{max} (see Figure 8.16(d-e)), a similar behavior is identified as in the classical regime, although the energy is more homogeneously distributed over all scales. At the height of the vortex inflow as seen in Figure 8.16(d), a large-scale peak is present, which shifts to $k_\varphi d \approx 3.9$ at $\tilde{r} = 0.1$. At the vortex outflow in Figure 8.16(f), this shift appears close to the OC wall at $\tilde{r} = 0.9$ for the same wavenumber. In the region of the vortex center (Figure 8.16(e)), the large-scale peak is most pronounced at $\tilde{r} = 0.5$ at a smaller wavenumber. For all the heights explored, the co-spectra do not reveal a peak in the small-scale regime. However, at the vortex inflow the energy is redistributed continuously from large to small scales, when $\tilde{r}$ is varied from 0.1 to 0.9, as it is clearly visible in the marked regime of $k_\varphi d \in [10, 20]$. It is also observed that the energy contained in the small scales increases in the vortex outflow region with decreasing radial position.

In summary, should prominent turbulent Taylor rolls exist in the flow, a peak at large scales exists and small-scale structures are present throughout the gap, but most prominently in the vortex ejecting regions. However, when the Taylor rolls have faded, the peak at large scales vanishes and small-scale structures are only detectable close to the cylinder walls. These findings reveal yet another evidence of the existence of turbulent non-axisymmetric Taylor rolls and support the idea that the large-scale rolls consist of small-scale unmixed plumes. Furthermore, the azimuthal length scale of these plumes is of the order of $k_\varphi d \in [10, 20]$.

8.5.2 Spatial correlation coefficients

As was shown in the previous section, the pre-multiplied energy co-spectra demonstrate the existence of small-scale structures (coherent plumes) and large-scale azimuthal structures, which are both connected due to the presence of turbulent Taylor vortices. For the sake of clarity, the investigation was confined to three specific cases. Within this section, however, the analysis of the large-scale peak is extended based on the azimuthal two-point auto-correlation coefficient. More specifically, the shift of the large-scale peak close to the cylinder walls as well as its characteristic in the center of the gap are worked out in more detail. The spatial two-point auto-correlation coefficient of the velocity fluctuations between two points, separated by $\Delta\varphi$ in the azimuthal direction, is given by [29]

$$R_{rr}\left(r, z, \Delta\varphi\right) = \frac{\langle u'_r\left(r, \varphi, z, t\right) u'_r\left(r, \varphi + \Delta\varphi, z, t\right)\rangle_{\varphi,t}}{\langle u'^2_r\left(r, \varphi, z, t\right)\rangle_{\varphi,t}}, \tag{8.9}$$

$$R_{\varphi\varphi}\left(r, z, \Delta\varphi\right) = \frac{\langle u'_\varphi\left(r, \varphi, z, t\right) u'_\varphi\left(r, \varphi + \Delta\varphi, z, t\right)\rangle_{\varphi,t}}{\langle u'^2_\varphi\left(r, \varphi, z, t\right)\rangle_{\varphi,t}}. \tag{8.10}$$

In Figure 8.17(a), the spatial two-point auto-correlation function of u_r is shown at $\tilde{r} = 0.1$ in the ultimate regime for three different flow cases at μ_{max}, and for both the vortex inflow and outflow. It can be observed that independently of the flow case, the auto-correlation function at the vortex inflow shows a minimum at $r\varphi/d \approx 0.13$, which indicates the existence of azimuthal structures of that size. In contrast, for the outflow a monotonic decrease of the auto-correlation function is observed. Close to the outer cylinder wall (see Figure 8.17(b)), the opposite behavior is observed with a pronounced minimum of the auto-correlation function for the vortex outflow at $r\varphi/d \approx 0.15$. This suggests that when the large-scale Taylor rolls transport fluid against the cylinder walls, azimuthal structures seem to be stimulated close to these walls, in good agreement with the findings of the spectral analysis shown in section 8.5.1. Next, in Figure 8.17(c,d), both the azimuthal and radial two-point auto-correlation function are shown for the center of the vortex evaluated at $\tilde{r} = 0.5$. At this position, the large-scale peak in the co-spectra

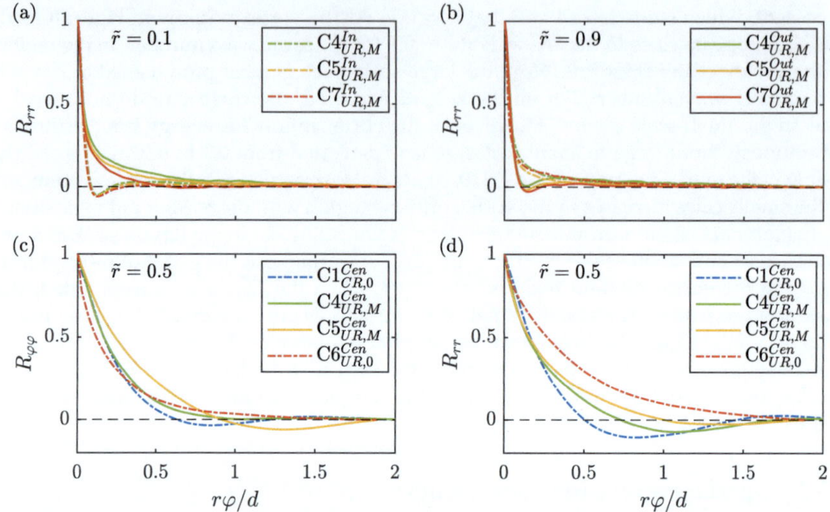

FIGURE 8.17: Upper row: Azimuthal two-point auto-correlation function of the fluctuation radial velocity component at the axial height of the vortex in- and outflow at (a) $\tilde{r} = 0.1$ and (b) $\tilde{r} = 0.9$. Lower row: Azimuthal two-point auto-correlation function of the fluctuation (c) azimuthal and (d) radial velocity component at the axial height of the vortex center at $\tilde{r} = 0.5$.

is most pronounced as it was shown in the Figures 8.15 and 8.16. Here, three flow states are compared in the classical and ultimate regime, where turbulent Taylor vortices are present, and a case, where they are absent (C6$_{UR,0}$). These figures reveal that in all vortex dominated cases, a large-scale oscillating behavior is developed, while in the absence of rolls, the auto-correlation simply decreases towards zero over the whole azimuthal measurement length. Thus, it can be concluded that the large-scale peak in the spectra apparently results from a wavy azimuthal pattern connected to the Taylor rolls.

8.6   Complex proper orthogonal decomposition (CPOD)

8.6.1   CPOD method

In order to reveal the flow structure connected to the large-scale oscillation within the flow found in figures 8.15, 8.16 and 8.17, a proper orthogonal decomposition (POD) of the velocity fluctuations field is performed. The POD, also known as empirical orthogonal function analysis, is a technique to extract modes like coherent structures from a flow field that contribute most to the energy of the flow [117]. It can be further used to identify the most relevant degrees of freedom in a dynamical system. For additional information, the reader is referred to the review of Berkooz *et al.* [10]. As a classical POD mode cannot capture propagating structures, a complex POD is performed, using the Hilbert transform (see Pfeffer *et al.* [92]), where a mode is split into two patterns, namely its real ($\Re$) and its imaginary part ($\Im$) with a phase difference of $\pi/2$, based on the algorithm described by

Marple [73]. Further, a combined analysis of both velocity components u_r and u_φ is used for the CPOD analysis. The applied algorithm is based on three steps [53]. At first, the Hilbert transform $\mathcal{H}$ is used to make the velocity components complex:

$$u'_{r,c} = u'_r + i\mathcal{H}(u'_r), \tag{8.11}$$
$$u'_{\varphi,c} = u'_\varphi + i\mathcal{H}(u'_\varphi). \tag{8.12}$$

Secondly, the imaginary, transformed velocity fluctuations $(u'_{r,c}, u'_{\varphi,c})$ are arranged in a data matrix $\mathbf{D}$, where columns are assigned to the spatial grid points $(x_1 : x_M)$ and rows to the time signals $(t_1 : t_N)$, while $M > N$:

$$\mathbf{D} = \begin{bmatrix} u'_{r,c}(x_1,t_1) & \cdots & u'_{r,c}(x_1,t_N) \\ \vdots & \ddots & \vdots \\ u'_{r,c}(x_M,t_1) & \cdots & u'_{r,c}(x_M,t_N) \\ u'_{\varphi,c}(x_1,t_1) & \cdots & u'_{\varphi,c}(x_1,t_N) \\ \vdots & \ddots & \vdots \\ u'_{\varphi,c}(x_M,t_1) & \cdots & u'_{\varphi,c}(x_M,t_N) \end{bmatrix}. \tag{8.13}$$

At the end, a singular value decomposition (SVD) of the data matrix with $\mathbf{D} = \mathbf{\Phi}\mathbf{\Sigma}\mathbf{\Psi}^T$ is performed. The columns of the matrix $\mathbf{\Phi}$ contain the left singular vectors of $\mathbf{D}$ and therefore the N complex modes $\mathbf{CPOD}(r, \varphi)$. The diagonal elements of $\mathbf{\Sigma}$ hold the singular values, whose square is a measure of the kinetic energy captured by the individual modes. Note that the singular values are sorted in descending order together with the complex modes, meaning that the first mode represents a larger fraction of kinetic fluctuation energy of the full field than the second one and so forth. Next, the velocity fluctuations at a specific axial location z can be reconstructed by

$$\mathbf{u}'(r, \varphi, t) = \sum_{i=1}^{N} a_i(t)\mathbf{CPOD}_i(r, \varphi). \tag{8.14}$$

The time-dependent coefficients $a_i(t)$ result from a projection of the complex modes onto the data matrix, i.e. $a_i = \mathbf{D}^T\mathbf{CPOD}_i$. In the following, the CPOD results are presented for two flow states at the height of the vortex center, where the azimuthal two-point autocorrelation showed a large-scale oscillating behavior. The first case is in the classical regime at $\mu = 0$ and the second one in the ultimate regime at μ_{max}.

8.6.2 CPOD in the classical regime

The first CPOD mode for $Re_S = 9.32 \times 10^3$ and $\mu = 0$ (C1$_{\text{CR,0}}$) is depicted in Figure 8.18 for both velocity components. With the first temporal coefficient $a_1(t)$ and the first mode $\mathbf{CPOD}_1$, the corresponding velocity fluctuation field $\mathbf{u}'_1(r, \varphi, t) = a_1(t)\mathbf{CPOD}_1(r, \varphi)$ is reconstructed. In Figure 8.18(a), the real part of $\mathbf{CPOD}_1$, representing the radial velocity component, shows nearly circular regions of positive and negative velocity, alternating in the azimuthal coordinate direction. This azimuthal wave pattern becomes azimuthally

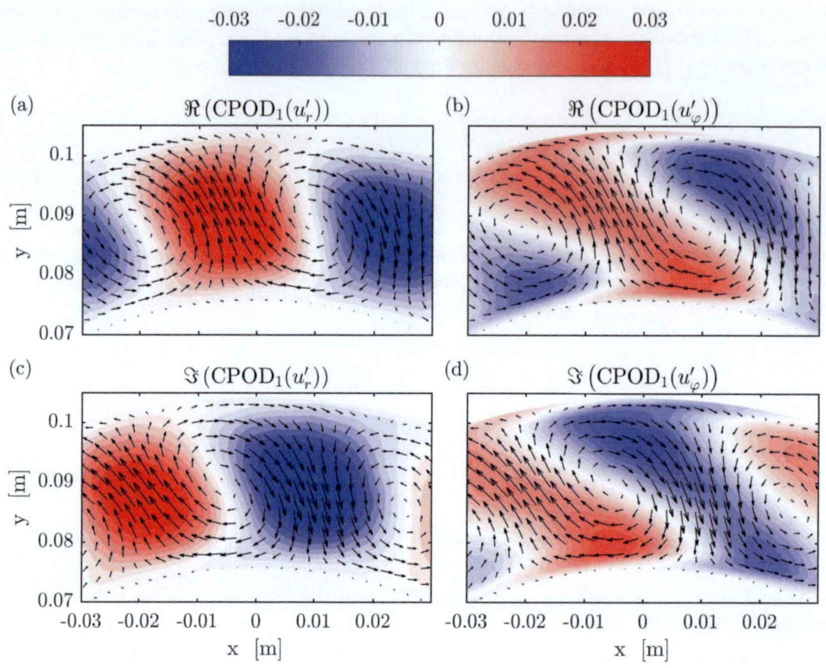

FIGURE 8.18: **CPOD$_1$** in the classical regime for $Re_S = 9.32 \times 10^3$ and $\mu = 0$ (C1$_{\mathrm{CR,0}}$) at the axial height of the vortex center for both velocity components. (a) Real and (c) imaginary part of **CPOD$_1$** for the radial velocity component. (b) Real and (d) imaginary part of **CPOD$_1$** for the azimuthal velocity component. Black arrows represent the resulting velocity field of **CPOD$_1$**.

shifted in the corresponding imaginary part in Figure 8.18(c), indicating an azimuthally traveling wave. The real part of the azimuthal velocity of **CPOD$_1$** in Figure 8.18(b) depicts diagonal bands with pointy edges close to the cylinder walls. These regions again show alternating positive and negative velocities. The corresponding imaginary part in Figure 8.18(d) is also azimuthally shifted, showing the same pattern. As a result, it is revealed that the velocity field consists of counter-rotating vortices in the horizontal plane, whose vorticity axes are coaxial to the rotation axis of the system. The pattern appears similar to that of the well-known wavy Taylor vortices at lower Re_S TC flow. Thus, here it is shown that in the classical turbulent regime, turbulent Taylor vortices can also feature azimuthal waves.

In Figure 8.19(a), the energy fractions captured by the CPOD modes are shown for $Re_S = 9.32 \times 10^3$ and $\mu = 0$ (C1$_{\mathrm{CR,0}}$). The first mode captures approximately 17% of the total energy fluctuation and a strong drop is observed to the second mode down to 6.7%. Accordingly, the first CPOD mode is dominant and is ,therefore, the only one discussed. The temporal energy spectrum E_t of the real part of the first mode's temporal coefficient $a_1(t)$ is pictured in Figure 8.19(b), together with the temporal energy spectrum of the full

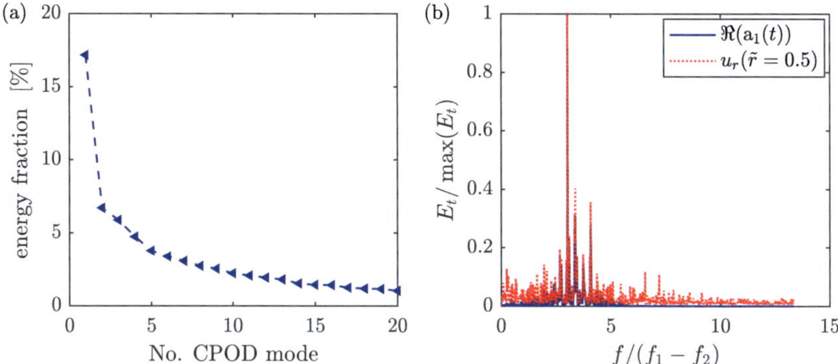

FIGURE 8.19: (a) Turbulent energy fraction captured by the CPOD modes for the case $\mu = 0$ and $Re_S = 9.32 \times 10^3$ (C1$_{\text{CR},0}$) in the classical regime at the axial height of the vortex center. Only the first 20 of the total 1500 modes are plotted. (b) Temporal power spectrum of the real part of the temporal coefficient $a_1(t)$ of **CPOD**$_1$ and of the radial velocity component of the full field at $\tilde{r} = 0.5$, averaged over φ. The frequency f is normalized by the difference of the cylinder frequencies $f_1 - f_2$.

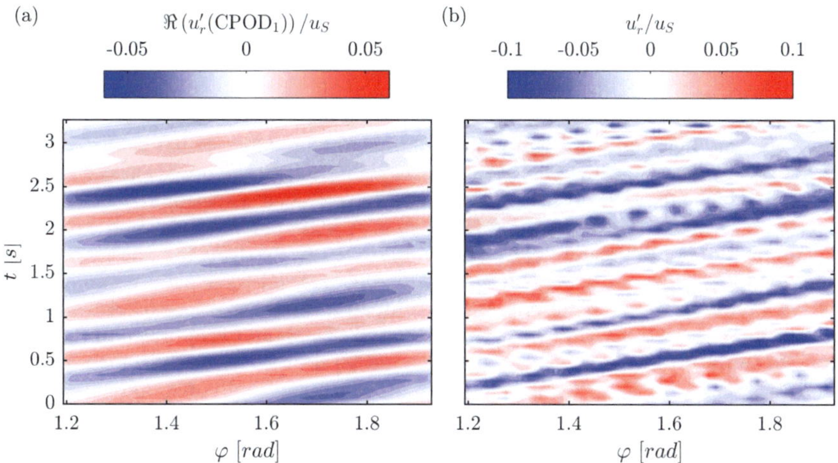

FIGURE 8.20: Space-time diagrams of (a) the reconstructed radial velocity component based on the first CPOD mode and (b) the full field radial velocity component for $Re_S = 9.32 \times 10^3$ and $\mu = 0$ (C1$_{\text{CR},0}$) in the classical regime at the axial height of the vortex center as a function of φ and t. The radial coordinate is fixed at $\tilde{r} = 0.5$.

field radial velocity component, calculated at $\tilde{r} = 0.5$ and averaged over φ. The dominant frequency of **CPOD**$_1$ is $f/(f_1 - f_2) = 3.03$, identical to that of the full field radial velocity component. Thus, it can be assumed that the first mode captures the temporal behavior of the full field quite accurately. The frequencies f_1 and f_2 represent the rotation frequencies of the inner and outer cylinder, respectively.

To illustrate the spatio-temporal character of the first mode, its reconstructed radial velocity component, as well as the corresponding full field, are depicted in Figure 8.20 at $\tilde{r} = 0.5$ as a function of the azimuthal coordinate φ and time t. The space-time diagram of $\mathbf{CPOD}_1$ depicts diagonal bands of alternating positive and negative velocity, representing azimuthally propagating waves into the direction of the mean flow. Furthermore, the same diagonal bands, superimposed on turbulent fluctuations, are observed in the full field. A reappearance of azimuthal waves for pure inner cylinder rotation at a similar radius ratio of $\eta = 0.733$ has already been found by Wang et al. [136] in the Reynolds number regime of $20 \leq Re/Re_{crit} \leq 38$. In the present study in the classical regime for $\mu = 0$ at $\eta = 0.714$, $Re_S/Re_{S,crit} \approx 99$ is found, where the critical shear Reynolds number is located around $Re_{S,crit} \approx 94.5$ [86]. This provides evidence of the reappearance of an azimuthal wave in the classical turbulent regime for a much larger turbulence intensity. Moreover, the corresponding wave frequency $f_{w,i}$, wavelength $\lambda_{w,i}$ and phase speed $c_{w,i}$ of $\mathbf{CPOD}_i$ can be extracted. The temporal phase function ϕ_i is defined as

$$\phi_i(t) = \arctan\left(\frac{\Im(a_i(t))}{\Re(a_i(t))}\right). \tag{8.15}$$

Its temporal derivative is equal to the angular frequency of the CPOD mode [115], leading to a wave frequency of $f_{w,1} = \partial_t \phi_1(t)/(2\pi) = 1.80\,\text{Hz}$. In addition, the azimuthal derivative of the spatial phase function Φ_i is defined as

$$\Phi_i(r, \varphi) = \arctan\left(\frac{\Im(\mathrm{CPOD}_i(r, \varphi))}{\Re(\mathrm{CPOD}_i(r, \varphi))}\right), \tag{8.16}$$

whose azimuthal derivative a measure for the local wavenumber $k_{\varphi,w,i}$. Note that the CPOD analysis yields only one temporal coefficient $a_i(t)$, but two spatial modes concerning the radial and azimuthal velocity components. Therefore two wavelengths can be extracted, which are however nearly identical. The averaged wavelength, evaluated for $\tilde{r} = 0.5$, is given by $\lambda_{w,1} = 2\pi/(\partial_\varphi \Phi_1(\tilde{r} = 0.5, \varphi)) = 0.68\,\text{rad}$, leading to an azimuthal wavenumber of approximately $k_{\varphi,w,1} = 2\pi/\lambda_{w,1} \approx 9$. In this way, the phase speed becomes $c_{w,1} = \lambda_{w,1}f_{w,1} = 1.22\,\text{rad/s} = 0.35\omega_1$. As the detected wave pattern reminds one of wavy Taylor vortices, the present result is compared to the wave speeds for wavy Taylor vortices measured by King et al. [65], although one should keep in mind the large Reynolds number difference. King et al. [65] found for the case of pure inner cylinder rotation at $\eta = 0.73$, $Re/Re_{crit} \approx 14$ and an average axial wavelength of $\lambda_w/d = 2.4$ a wave speed of $c_w \approx 0.2\omega_1$, while the corresponding pattern featured $k_{\varphi,w} = 2$. Additionally, they showed for a slightly larger radius ratio of $\eta = 0.84$, that the wave speed increases when $Re/Re_{crit} > 18$. Thus, it can be concluded that the wave speeds of the azimuthally traveling waves, which are superimposed on the Taylor vortices, are of the same order in the wavy Taylor vortex regime and in the turbulent Taylor vortex regime. Moreover, the wave speed seems to increase with the forcing Re_S, as the wave speed in the present case is noticeably higher than the one reported by King et al. [65].

8.6.3 CPOD in the ultimate regime

Now, the focus is turned to the ultimate regime at the torque maximum rotation rate. In Figure 8.21(a-d), the first CPOD mode, representing the real and imaginary parts of the radial and azimuthal velocity component, is depicted for $Re_S = 6.68 \times 10^4$ and $\mu = -0.36$ (C4$_{\text{UR,M}}$) at the axial height of the vortex center. The flow in the ultimate regime features nearly the same pattern as already shown in 8.18 with alternating regions of positive and negative velocity. In the case of the radial and azimuthal velocity components, these regions form circular patches and diagonal bands with pointy edges, respectively. In addition, the patterns of the imaginary part are azimuthally shifted relative to the real parts.

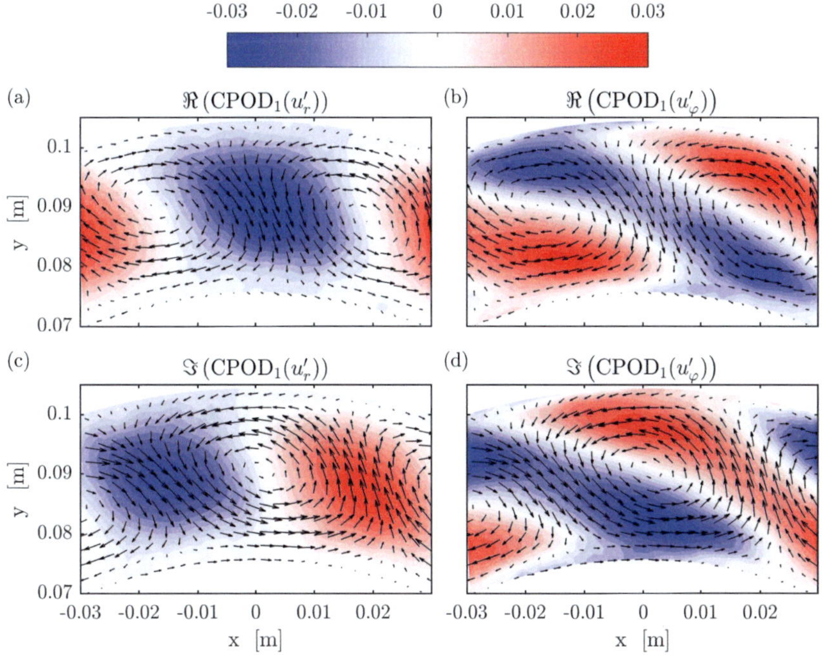

FIGURE 8.21: **CPOD$_1$** in the ultimate regime for $Re_S = 6.68 \times 10^4$ and $\mu = -0.36$ (C4$_{\text{UR,M}}$) at the axial height of the vortex center for both velocity components. (a) Real and (c) imaginary part of **CPOD$_1$** for the radial velocity component. (b) Real and (d) imaginary part of **CPOD$_1$** for the azimuthal velocity component. Black arrows represent the resulting velocity field of **CPOD$_1$**.

The turbulent energy fractions in Figure 8.22(a) show similar to the previous case in the classical regime a strong drop between the first and second mode from approximately 12 % to 5 %. Further, the temporal power spectrum of the temporal coefficient $\Re(a_1(t))$ and of the full field in terms of u_r at $\tilde{r} = 0.5$ depict again a prominent peak in Figure 8.22(b) at $f/(f_1 - f_2) = 0.92$. It is worth mentioning, that this peak is much more prominent than the one seen in the classical regime without the emergence of any secondary peak or higher harmonics.

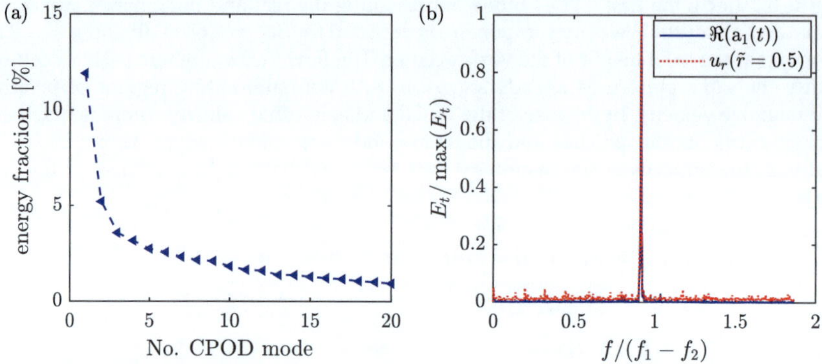

FIGURE 8.22: (a) Turbulent energy fraction captured by the CPOD modes for the case $\mu = -0.36$ and $Re_S = 6.68 \times 10^4$ (C4$_{\mathrm{UR,M}}$) in the ultimate regime at the axial height of the vortex center. Only the first 20 of the total 1500 modes are plotted. (b) Temporal power spectrum of the real part of the temporal coefficient $a_1(t)$ of **CPOD$_1$** and of the radial velocity component of the full field at $\tilde{r} = 0.5$, averaged over φ. The frequency f is normalized by the difference of the cylinder frequencies $f_1 - f_2$

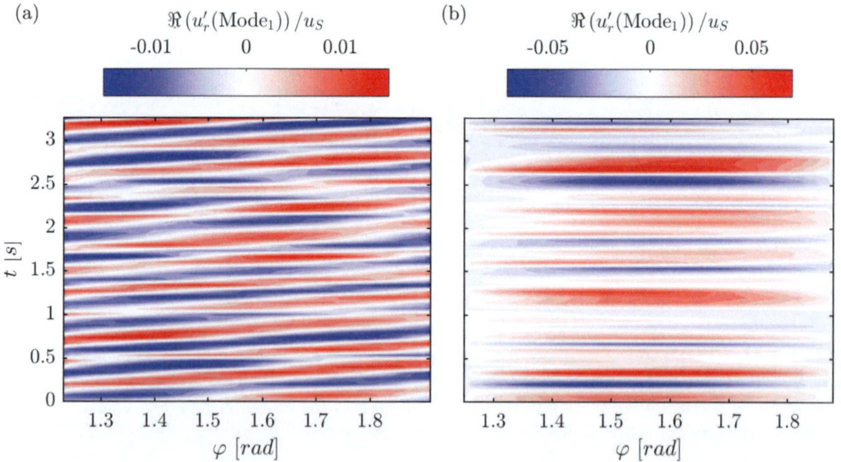

FIGURE 8.23: Space-time diagrams of the reconstructed radial velocity component based on the first CPOD mode for (a) $Re_S = 6.69 \times 10^4$ and $\mu = -0.36$ (C4$_{\mathrm{UR,M}}$) and (b) $Re_S = 2.15 \times 10^5$ and $\mu = 0$ (C6$_{\mathrm{UR,0}}$) at the axial height of the vortex center as function of φ and t. The radial coordinate is fixed at $\tilde{r} = 0.5$.

In Figure 8.23(a) the space-time diagram of the reconstructed radial velocity component based on $\mathbf{CPOD}_1$ depicts the before seen diagonal bands of alternating velocity as result of azimuthally traveling waves. In comparison to C1$_{\mathrm{CR},0}$, the frequency of the wave has strongly increased, as visible by the higher number of bands within the same time interval. The wave frequency becomes $f_{w,1} = 3.67\,\mathrm{Hz}$, the wavelength $\lambda_{w,1} = 0.74\,\mathrm{rad}$ and the phase speed $c_{w,1} = 2.72\,\mathrm{rad/s} = 0.11\Delta\omega$. Remarkably, the turbulent Taylor vortices feature azimuthal waves also in the ultimate turbulent regime at μ_{max}. To the best of our knowledge, this is the first study reporting such a wave pattern in that highly turbulent regime.

As an additional test, a CPOD analysis is performed for $Re_S = 2.15 \times 10^5$ and $\mu = 0$ (C6$_{\mathrm{UR},0}$) in the ultimate regime, where the large-scale Taylor rolls disappear in the mean field. The space-time diagram for the reconstructed radial velocity component connected to the first CPOD mode is depicted in Figure 8.23(b). The diagram shows horizontal bands with a random size. Therefore, no traveling waves can be detected, when large-scale Taylor rolls are absent in the very high ultimate regime at $\mu = 0$. As a comparison tool, the results concerning the azimuthally traveling waves found in this study and the findings of King $et\ al.$ [65] for $\eta = 0.73$ are shown in table 8.2.

	η	Re_S	μ	$k_{\varphi,w,1}$	$c_{w,1}\ [\mathrm{rad/s}]$
Classical regime (C1$_{\mathrm{CR},0}$)	0.714	9.32×10^3	0	≈ 9	$0.35\Delta\omega$
Ultimate regime (C4$_{\mathrm{UR},M}$)	0.714	6.68×10^4	-0.36	≈ 9	$0.11\Delta\omega$
King $et\ al.$ [65]	0.730	1.05×10^3	0	2	$0.21\Delta\omega$

TABLE 8.2: Overview of detected azimuthally traveling waves in the classical and ultimate regime in comparison to the findings of King $et\ al.$ [65].

To what extent the detection of azimuthally traveling waves is dependent on the turbulence level (Re_S) and the geometry of the system (η, Γ) is not known. It has to be noted that further experimental as well as numerical investigations are required to address this issue and reveal whether it is an intrinsic property of the flow or not.

8.7 Conclusion

By means of planar PIV measurements performed in horizontal planes at different axial heights, the dependence of the small-scale statistics and flow organization on the presence of large-scale Taylor rolls is shown in high Reynolds number TC flow. The ratio of angular velocities μ is the appropriate parameter to control whether or not the Taylor rolls within the gap are prominent and stable in a statistical sense. While in the classical turbulent regime Taylor rolls are observed for both $\mu = 0$ and $\mu < 0$, they fade out when the ultimate regime is reached for $\mu = 0$, but can be formed again when sufficient counter-rotation ($\mu < 0$) is introduced, in particular at μ_{max} where the flux of angular momentum is maximal. These turbulent states can then be used to compare the flow, for the same Re_S, with and without the presence of these large-scale rolls.

To uncover the interplay between large-scale rolls and small-scales plumes in a statistical manner, PDFs of both velocity components and the net convective Nusselt number have been evaluated over cylindrical surfaces and at specific axial positions. The PDFs of the radial and azimuthal velocity component are close to Gaussian, when evaluated at a fixed height, in accordance with the results of Huisman *et al.* [58]. However, when all heights covering one vortex pair are included, the Gaussian shape is only preserved for $\mu = 0$ for both velocity components but changes drastically for μ_{max}. There, the PDFs of u_φ feature a cusp-like shape with exponential tails, which is a fingerprint of intermittent small-scale plumes and can also be found in RB flow concerning the temperature [16, 35]. The PDFs of $\mathrm{Nu}_\omega^{c,net}$ fluctuate around zero and not the averaged value of the Nusselt number and can be described (for $\mu = 0$) by the shape of a distribution of the product of two Gaussian variables. At μ_{max} however, the right-hand tail broadens with increasing Re_S, which can be attributed to an increasing number of rare and strong events of momentum flux at the axial location of the vortex in- and outflow. These axially dependent events originate from the ejection of coherent plumes from the cylinder walls into the gap and lead to a dominant contribution of the vortex in- and outflow to the overall momentum transport. While the contribution of the vortex inflow reaches a nearly fixed value in the ultimate regime of $\approx 30\%$ at μ_{max} in the outer gap region, the contribution of the outflow increases monotonically with Re_S. Here, a strikingly large value of the angular momentum transport of $\approx 60\%$ is found in the inner gap region at $Re_S = 2.14 \times 10^5$.

The energy content and azimuthal length scale of these small-scale flow structures in the presence of large-scale rolls are calculated based on azimuthal energy co-spectra. The small-scale plumes show an azimuthal extent of $k_\varphi d \in [10, 20]$ and contribute most to the overall fluctuation energy in regions where the large-scale rolls transport fluid away from the wall, i.e. the so-called ejection regions [86]. In addition, the energy spectra combined with a correlation analysis revealed that the large-scale Taylor rolls feature an azimuthally oscillating behavior instead of being axisymmetric and stimulate mid-scale azimuthal structures of sizes $r\varphi/d \approx 0.13 - 0.15$, where the fluid transported by these rolls impacts on the cylinder walls.

In order to finally capture the underlying flow structure of non-axisymmetric Taylor rolls, a complex POD analysis was performed at the height of the vortex center. Large-scale azimuthally traveling waves are superimposed on the turbulent Taylor vortices, not only in the classical regime at $\mu = 0$, but also in the ultimate regime at μ_{max}. These waves propagate into the direction of the mean flow and are similar to the well-known wavy Taylor vortices. While the wave speeds of $0.35\Delta\omega$ in the classical and $0.11\Delta\omega$ in the ultimate regime are of the same order as the one for the wavy Taylor vortices measured by King *et al.* [65], the azimuthal wave number found, of $k_{\varphi,w,1} = 9$, is much higher than in the laminar regime.

These findings reveal the intrinsic statistical relation between structures of different scales: large-scale Taylor rolls and small-scale plumes in high Reynolds number TC flow. Their interplay strongly relies on the specific locations along the vortex (inflow, center or outflow) and since they play a prominent role for the flow organization and the momentum transport, the present study underlines the importance of an axial exploration when studying the statistics of turbulent TC flows.

It can be finally noted that we believe that the present results are much more general than only holding for turbulent TC flow. Clearly, they will generalize to turbulent RB flow with its organization on very large scales [89, 114], but also to pipe and channel flow with highly organized structures in spanwise direction [61, 75, 112]. What mechanism however sets the length scale of the organization of such superstructures in the spanwise direction remains unclear.

Chapter 9

Summary & conclusion

In this thesis, various aspects of the angular momentum transport in fully turbulent Taylor-Couette flow have been experimentally investigated for three different radius ratios. As the transport is strongly influenced by the local and global flow organization, direct torque measurements are complemented by flow visualizations and PIV measurements in horizontal planes at different cylinder heights. The focus was on the investigation of the global momentum transport within the unexplored wide-gap configuration of $\eta = 0.357$, the influence of large-scale turbulent Taylor vortices on the global velocity statistics and their contribution to the overall momentum transport for $\eta = 0.5$, and the interaction of these Taylor vortices with small-scale plumes for $\eta = 0.714$. The angular momentum transport features a maximum at a constant shear rate as a function of rotation for all three η-configurations with the same physical mechanism of strengthened turbulent Taylor vortices. Therefore, it can be stressed that the presented findings in the different radius ratio experiments reveal universal insights into the transport process for this η-regime.

After introducing the essential fluid dynamical fundamentals for this study in Chapter 2, the actual state of research in fully turbulent TC flow is discussed in Chapter 3, concerning the momentum transport. In Chapter 4 the experimental facilities used within this thesis are presented. The TvTCC experiment is characterized by full optical access to the flow through a transparent outer cylinder and top plate, a torque sensor unit integrated into the inner cylinder power train and an exchangeable inner cylinder to realize different radius ratios η. The cylinder speeds are controlled with an absolute accuracy of the mean of 0.6 % and fluctuations smaller than 1 %. Further, the fluid temperature increase due to friction losses in the gap is monitored by a thermocouple type K sensor with an accuracy of 0.8 % and the actual temperature dependent fluid properties (density ρ and kinematic viscosity ν) are corrected based on viscosimetry measurements. In the TvTCC experiment torque measurements and flow visualizations were performed for $\eta = 0.357$ and PIV measurements in horizontal planes at different cylinder heights for $\eta = 0.5$. In the BTTC experiment in Twente, whose properties have been described in detail by Huisman *et al.* [59], similar PIV measurements were carried out for a radius ratio of $\eta = 0.714$.

The used measurement techniques are described explicitly in Chapter 5. For the purpose of flow visualization, the silicone oil M3 is seeded with elliptically shaped aluminum flake particles and blue pigments, to reveal the flow organization in a thin layer close to the outer cylinder wall. The light of halogen lamps is reflected with different intensities according to the flake orientation in the flow, and captured by a high-speed camera. After correction for inhomogeneous illumination by a calibration measurement with a stable flow state, space-time diagrams representing the intensity information of one axial line over time are used to uncover large-scale flow structures. Torque measurements are realized at the inner cylinder based on a contactless shaft-to-shaft rotary torque sensor, where friction forces of the fluid over the whole cylinder length and those of the IC bearings are included. Because of the superimposed bearing friction, the torque is measured twice, once with a liquid working fluid and once with air. Subtraction of both signals yields the pure friction force of the fluid. The torque signal has been validated for $\eta = 0.5$ by comparison with direct torque measurements and numerical simulations of Merbold *et al.* [78] and Ostilla-Mónico *et al.* [87], showing a very good agreement especially for Reynolds numbers smaller than $Re_S < 10^5$, with a relative deviation smaller than 5 %. Further, a minimum measurement time of 70 s is needed to converge the progressing average of the torque signal to a constant value. To measure the velocity field in horizontal planes at different cylinder heights, a double-pulsed cavity laser is mounted on a traverse, creating a thin light sheet in the horizontal plane. The fluid is seeded with spherical tracers and their reflected light is captured by a CCD Camera focused on the light sheet through the transparent top plate. The radial and azimuthal velocity components are calculated based on a multi-pass cross-correlation algorithm, and the calibration from image to real coordinates is performed by fitting two concentric circles to the cylinder walls on the raw intensity images. To quantify the accuracy of the PIV measurements, the magnification or scale factor between image and real coordinates as a function of the axial height of the light sheet is evaluated, finding a nearly linear relationship with a correlation coefficient of 0.9996. This relationship demonstrates the accurate alignment of the PIV setup. Further, the velocity field is measured for solid body rotation, which yields the laminar Couette solution in the gap. The relative deviation of the measured azimuthal velocity component from the laminar Couette solution does not exceed 0.8 %, which quantifies the absolute measurement error.

With the proof of the functional capability of the used experimental facilities and measurement techniques, the scientific questions raised in Chapter 1 could be answered. Chapter 6 dealt with the effective scaling of the angular momentum transport with shear (Re_S) and rotation (μ) for a wide gap of $\eta = 0.357$. For pure inner cylinder rotation the Nusselt number Nu_ω increases with Re_S and the local effective scaling exponent α, based on a power-law ansatz of the form $\mathrm{Nu}_\omega \sim Re_S^{\alpha-1}$, revealed a transition in the range of $1.3 \times 10^4 \leq Re_S \leq 4 \times 10^4$. Contrary to the first appearance of this behavior as boundary layer transition that triggers the ultimate regime, the investigations here can be assigned to the classical regime. This is obvious since the effective scaling exponent α is smaller than the predicted limit of $\alpha_{marg} = 5/3$ and the outer boundary layer does not feature an extended logarithmic region. Instead, the observed transition can be associated with the capacity of the outer cylinder to emit small-scale plumes at a rate that equalize the angular momentum in the bulk flow. While the frictional Reynolds number at the inner cylinder approaches to $Re_{\tau,i} \approx 230$ at the beginning of the transition, the outer frictional Reynolds number reaches the same value at its end. Accordingly, the effects of concavity and curvature become very pronounced in case of wide-gap TC flows and cause a delayed transition at the low-sheared and stabilized outer cylinder. For independently rotating cylinders at a constant shear rate, a maximum appears in the Nusselt number for slight counter-rotation at $\mu_{max} = -0.123 \pm 0.030$ due to the formation of large-scale turbulent Taylor rolls in the gap. The prediction of the location of μ_{max} according to Brauckmann and Eckhardt [15] is within the calculated uncertainty interval. As that prediction also holds for radius ratios of $\eta = 0.5$ and $\eta = 0.714$ [15, 87], it can be stressed that the physical mechanism of the maximum momentum transport is independent of the curvature in terms of η within the investigated radius ratio regime. Further, two different axial wavelengths for the turbulent Taylor rolls ($\lambda_{TV}(Re_S = 2.5 \times 10^4) = 1.02d$ and $\lambda_{TV}(Re_S = 5.0 \times 10^4) = 0.73d$ at μ_{max}) could be observed, a feature that is also shared with smaller gap TC flows [57, 127]. By comparison of torque data for $\eta = 0.357$ and $\eta = 0.5$, it could be shown that curvature affects the momentum transport only for rotation ratios smaller than μ_{max}, where a radial partitioning of the flow occurs (see also Brauckmann *et al.* [17] and Dubrulle *et al.* [31]).

The influence of the large-scale Taylor rolls onto the global velocity statistics and their contribution to the overall angular momentum transport was analyzed in Chapter 7 for $\eta = 0.5$. Here, two μ-curves for $Re_S = 8 \times 10^4$ and $Re_S = 10^5$ are analyzed, which show the same characteristics. At $\mu_{max}(\eta = 0.5) = -0.2$, the rotation ratio of the torque maximum, pronounced large-scale Taylor rolls are found in the mean field identical to the case of $\eta = 0.357$. There, the radial profiles of the angular velocity $\langle \tilde{\omega}(r) \rangle_{\varphi,z,t}$ yield a nearly horizontal shape with a vanishing radial gradient. This shape reflects the effective mixing of angular momentum inside the gap at the maximum of torque. The same horizontal shape has also been found for smaller gaps [87]. Further, the neutral line r_n detaches from the outer cylinder wall, when μ becomes smaller than μ_{max}. This detachment is the beginning of the radial flow partitioning, where curvature effects begin to influence the momentum transport, as shown for $\eta = 0.357$. To analyze the process of strengthening and weakening of turbulent Taylor vortices, when Re_S is kept constant and μ is decreased, starting in the co-rotating regime, the turbulent and large-scale circulation kinetic energy as well as the wind Reynolds number have been calculated. The formation of the Taylor vortices starts around $\mu = 0$, and they are strengthened continuously with decreasing μ up to μ_{max}, while for more negative μ, the breakdown of the large-scale vortices occurs suddenly. Besides, the scaling of the wind Reynolds number with the shear Reynolds number for $\mu = 0$ is nearly identical to the proposed scaling in the RB flow, which is a clear evidence for the similarity of both systems. Up to this point, it could only be shown that the maximum angular momentum transport is connected to the large-scale Taylor rolls. To evaluate their importance for the transport, the Nusselt number has been decomposed into its large-scale and turbulent contributions. For $Re_S = 10^5$ and $\mu = \mu_{max}$, the momentum transport due to the Taylor vortices is more than twice as large as the contribution of the turbulent fluctuations and vanishes for $\mu \geq 0$. In comparison to the results of Brauckmann and Eckhardt [15] using the same decomposition at $Re_S = 2 \times 10^4$, the range of rotation ratios, where turbulent Taylor vortices can exist, shrinks strongly with increasing Re_S.

The interaction of the large-scale Taylor rolls and small-scale plumes concerning the momentum transport was investigated in Chapter 8. Here, flows with and without prominent Taylor rolls where compared by adjusting μ either to the rotation ratio of the torque maximum (μ_{max}) or pure inner cylinder rotation ($\mu = 0$). The PDFs of the azimuthal velocity component evaluated over cylindrical surfaces and time are nearly Gaussian for $\mu = 0$ and depict a cusp-like form with exponential tails for μ_{max}. The latter shape is identical to the one found for the temperature PDF in RB flow [17, 35], whose origin could be linked to the coexistence of bursting plumes and large-scale convection rolls. To uncover the momentum transport in more detail, the local net convective Nusselt number $Nu_\omega^{c,net}$ is evaluated. The momentum transport in the presence of Taylor rolls takes place mainly at the axial height of the vortex in- and especially outflow, where the radial and azimuthal velocity components are highly correlated. In the ultimate turbulent regime, the momentum transport at the vortex inflow is nearly constant and amounts to around 30 % of the overall transport in the outer gap region, while the contribution of the vortex outflow increases with Re_S and reaches an amount of 60 % for $Re_S = 2,14 \times 10^5$ in the inner gap region. The PDFs of $\mathrm{Nu}_\omega^{c,net}$ evaluated over cylindrical surfaces and time are highly skewed in the presence of Taylor vortices, which leads to extreme but rare events of momentum transport, connected again to the vortex in- and outflow. The same PDF shape of the heat transport appears in RB flow, where these rare and extreme events of heat transport are footprints of thermal plumes [109]. Based on the analogy between TC and RB flow, it can be concluded that plumes transporting angular momentum in TC flow are emitted strongly from the cylinder walls in the region of the vortex in- and outflow. To verify the existence of small-scale plumes, azimuthal energy co-spectra are calculated. There, a direct fingerprint of these plumes is found in form of a peak in the pre-multiplied spectra at small scales around $k_\varphi d \in [10,20]$. This peak is most pronounced at the axial height of the vortex outflow close to the inner cylinder wall as well as at the height of the vortex inflow close to the outer cylinder wall. Accordingly, the plume ejection regions coincide with locations where the turbulent Taylor vortices transport fluid away from the cylinder walls. In addition, a peak at large scales has been found in the spectra, which depicts an oscillating behavior according to a correlation analysis. This peak results from azimuthally traveling waves superimposed on the Taylor vortices that are similar to the well-known wavy Taylor vortex flow. To the best of our knowledge, these wave patterns have never been detected in the classical turbulent region for $\mu = 0$ at such high Re_S and in the ultimate turbulent region for μ_{max}.

To sum up, the angular momentum transport in medium and wide gaps with radius ratios of $(0.357 \leq \eta \leq 0.714)$ features a maximum in the slight counter-rotating regime (μ_{max}), caused by strengthened turbulent Taylor vortices. As the cylinder curvature effects the Nusselt number only for $\mu < \mu_{max}$, it can be assumed that the individual findings at μ_{max} for the three different radius ratios are valid for all these systems. At μ_{max} the momentum is mainly transported by the large-scale Taylor rolls and not turbulence. Further, this large-scale momentum transport is focused on the vortex in- and outflow regions, where small-scale momentum plumes are emitted by the cylinder walls, feeding the large-scale rolls. Finally, the highlighted similarities of the results of this thesis with findings in RB flow suggest that these results can be generalized to other turbulent flows.

Based on the findings of this thesis, three topics should be treated in the future. At first, this study illustrates general flow features that are shared between TC and RB flow, namely the interaction of large-scale rolls (either Taylor vortices or Bénard cells) and small-scale plumes (transporting either momentum or heat). Such highly-organized structures also can be found in pipe or channel flows [61, 112], and especially the analogy between TC and pipe flow concerning the momentum transport and dissipation rate according to [32] (see Section 2.6) calls for a combined analysis of these systems. Secondly, in this thesis, the dependence of the momentum transport on large- and small-scale structures in the fully turbulent regime could only be analyzed based on statistics. To investigate such coherent structures in more detail, measurement techniques should be used, that capture the instantaneous 3D velocity field. Despite the difficulties to realize such measurements in a turbulent TC flow (see Section 2.6), they are highly needed for a better understanding of the underlying flow structures. At last, TC investigations of even wider gaps than $\eta = 0.35$ are still missing. According to Figure 3.2, the effect of curvature increases strongly in that completely unexplored parameter regime, which may result in new flow states. Further, Brauckmann *et al.* [17] could show, that in case of very narrow gaps ($\eta > 0.8$), the momentum transport features a different behavior to what has been shown within this thesis. There, instead of one broad maximum, a narrow and a broad maximum occur, when Re_S is kept constant and only μ is changed. Obviously, the vanishing of curvature effects strongly influences the momentum transport. Accordingly, when curvature effects become more dominant in case of $\eta \to 0$, also prominent changes concerning the momentum transport could be possible. Within the actual configuration of the TvTCC experiment, radius ratios of $\eta = 0.2$ and $\eta = 0.1$ are accessible. Therefore, flow visualizations and PIV measurements could be realized right now. However, the actual torque measurement unit needs to be replaced. When the inner cylinder radius decreases, the friction force of the fluid decreases when a constant shear rate is assumed, because the surface that is in contact with the fluid is reduced. Therefore, the torque induced by the fluid will not be sufficiently larger than the one due to the friction of the bearings. Splitting the inner cylinder and measuring the torque only at a middle segment is also not appropriate, as the inner cylinder radii are very small and different inner cylinders shall be used. The best solution could be, to measure torque at the outer cylinder by constructing a double-walled OC connected via strain gauges.

Bibliography

[1]   N. Abcha, N. Latrache, F. Dumouchel, and I. Mutabazi. "Qualitative relation between reflected light intensity by Kalliroscope flakes and velocity field in the Couette-Taylor system." In: *Exp. Fluids* 45 (2008), pp. 85–94.

[2]   R. J. Adrian. "Twenty years of particle image velocimetry." In: *Exp. Fluids* 39 (2005), pp. 159–169.

[3]   C. D. Andereck, S. S. Liu, and H. L. Swinney. "Flow regimes in a circular Couette system with independently rotating cylinders." In: *J. Fluid Mech.* 164 (1986), pp. 155–83.

[4]   K. Avila and B. Hof. "High-precision Taylor-Couette experiment to study subcritical transitions and the role of boundary conditions and size effects." In: *Rev. of. Sc. Instr.* 84 (2013), p. 065106.

[5]   K. Avila, D. Moxey, A. de Lozar, M. Avila, D. Barkley, and B. Hof. "The Onset of Turbulence in Pipe Flow." In: *Science* 333 (2011), pp. 192–196.

[6]   M. Avila. "Stability and Angular-Momentum Transport of Fluid Flows between Corotating Cylinders." In: *Phys. Rev. Lett.* 108 (2012), p. 124501.

[7]   M. Avila, M. Grimes, J. M. Lopez, and F. Marques. "Global endwall effects on centrifugally stable flows." In: *Phys. Fluids* 20 (2008), p. 104104.

[8]   A. Barcilon and J. Brindley. "Organized structures in turbulent Taylor-Couette flow." In: *J. Fluid. Mech.* 143 (1984), pp. 429–449.

[9]   M. Le Bars and P. Le Gal. "Experimental analysis of the stratorotational instability in a cylindrical Couette flow." In: *Phys. Rev. Lett.* 99 (2007), p. 064502.

[10]   G. Berkooz, P. Holmes, and J. L. Lumley. "The Proper Orthogonal Decomposition in the Analysis of Turbulent Flows." In: *Annu. Rev. Fluid Mech.* 25 (1993), pp. 539–75.

[11]   H. Blasius. "Grenzschichten in Flüssigkeiten mit kleiner Reibung." In: *Zeitschrift fü Mathematik und Physik.* 56 (1908).

[12]   D. Borrero-Echeverry, M. F. Schatz, and R. Tagg. "Transient turbulence in Taylor-Couette flow." In: *Phys. Rev. E* 81 (2010), p. 025301.

[13]   Peter Bradshaw. *Effects of streamline curvature on turbulent flow.* Tech. rep. Advisory Group For Aerospace and Development, 1973.

[14]   H. J. Brauckmann and B. Eckhardt. "Direct numerical simulations of local and global torque in Taylor-Couette flow up to Re = 30000." In: *J. Fluid Mech.* 718 (2013a), pp. 398–427.

[15]   H. J. Brauckmann and B. Eckhardt. "Intermittent boundary layers and torque maxima in Taylor-Couette flow." In: *Phys. Rev. E* 87 (2013b), p. 033004.

[16] H. J. Brauckmann, B. Eckhardt, and J. Schumacher. "Heat transport in Rayleigh-Bénard convection and angular momentum transport in Taylor-Couette flow: a comparative study." In: *Phil. Trans. R. Soc. A* 375 (2016), p. 20160079.

[17] H. J. Brauckmann, M. Salewski, and B. Eckhardt. "Momentum transport in Taylor-Couette flow with vanishing curvature." In: *J. Fluid Mech.* 790 (2016), pp. 419–452.

[18] D. J. Burin and C. J. Czarnocki. "Subcritical transition and spiral turbulence in circular Couette flow." In: *J. Fluid Mech.* 709 (2012), pp. 106–122.

[19] M. J. Burin, E. Schartmann, and H. Ji. "Local measurements of turbulent angular momentum transport in circular Couette flow." In: *Exp. Fluids* 48 (2010), pp. 763–769.

[20] B. Castaing, G. Gunaratne, F. Heslot, L. Kadanoff, A. Libchaber, S. Thomae, X. Z. Wu, S. Zaleski, and G. Zanetti. "Scaling of hard thermal turbulence in Rayleigh-Bénard convection". In: *J. Fluid Mech.* 204 (1989), pp. 1–30.

[21] S. Chandrasekhar. *Hydrodynamic and Hydromagnetic Instability.* 1st ed. Oxford: Clarendon Press, 1961.

[22] C. R. Chu, M. B. Parlange, G. G. Katul, and J. D. Albertson. "Probability density functions of turbulent velocity and temperature in the atmospheric surface layer." In: *Water Resources Research* 32(6) (1996), pp. 1681–1688.

[23] D. Coles. "Transition in circular Couette flow." In: *J. Fluid Mech.* 21 (1965), pp. 385–425.

[24] M. M. Couette. "Études sur le frottement des liquides." In: *Ann. Chim. Phys.* 6 (1890), pp. 433–510.

[25] K. Coughlin and P. S. Marcus. "Turbulent bursts in Couette-Taylor flow." In: *Phys. Rev. Lett.* 77 (1996), pp. 2214–2217.

[26] R. C. DiPrima and H. L. Swinney. "Hydrodynamic Instabilities and the Transition to Turbulence." In: *Topics in Applied Physics.* 45 (1985), pp. 139–180.

[27] C. R. Doering and P. Constantin. "Energy dissipation in shear driven turbulence." In: *Phys. Rev. Lett.* 69 (1992), pp. 1648–1651.

[28] S. Dong. "Direct numerical simulation of turbulent Taylor-Couette flow." In: *J. Fluid Mech.* 587 (2007), pp. 373–93.

[29] S. Dong. "Turbulent flow between counter-rotating concentric cylinders: a direct numerical simulation study." In: *J. Fluid Mech.* 615 (2008), pp. 371–99.

[30] R. J. Donnelly. "Taylor-Couette flow: the early days." In: *Phys. Today.* 44 (1991), pp. 32–39.

[31] B. Dubrulle, O. Dauchot, F. Daviaud, P.-Y. Longaretti, D. Richard, and J.-P. Zahn. "Stability and turbulent transport in Taylor-Couette flow from analysis of experimental data." In: *Phys. Fluids* 17 (2005), p. 095103.

[32] B. Eckhardt, S. Grossmann, and D. Lohse. "Fluxes and energy dissipation in thermal convection and shear flows." In: *EPL (Europhysics Letters)* 78.2 (2007), p. 24001.

[33] B. Eckhardt, S. Grossmann, and D. Lohse. "Torque scaling in turbulent Taylor-Couette flow between independently rotating cylinders." In: *J. Fluid Mech.* 581 (2007b), pp. 221–250.

[34] B. Eckhardt, T. M. Schneider, B. Hof, and J. Westerweel. "Turbulence Transition in Pipe Flow." In: *Annu. Rev. Fluid Mech.* 39 (2007), pp. 447–68.

[35] M. S. Emran and J. Schumacher. "Fine-scale statistics of temperature and its derivatives in convective turbulence." In: *J. Fluid Mech.* 611 (2008), pp. 13–34.

[36] A. Esser and S. Grossmann. "Analytic expression for Taylor-Couette stability boundary." In: *Phys. Fluids* 8 (1996), pp. 1814–1819.

[37] R. Ezeta, S. G. Huisman, C. Sun, and D. Lohse. "Turbulence strength in ultimate Taylor-Couette turbulence." In: *J. Fluid Mech.* 836 (2017), pp. 397–412.

[38] H. Faisst and B. Eckhardt. "Transition from the Couette-Taylor system to the plane Couette system." In: *Phys. Rev. E* 61 (2000), pp. 7227–7230.

[39] M. A. Fardin, C. Perge, and N. Taberlet. "The hydrogen atom of fluid dynamics - introduction to the Taylor-Couette flow for soft matter scientists." In: *Soft Matt.* 10 (2014), pp. 3523–3535.

[40] P. Fenstermacher, H. Swinney, and J. Gollub. "Dynamical instabilities and the transition to chaotic Taylor vortex flow." In: *J. Fluid Mech.* 94(1) (1979), pp. 103–128.

[41] A. Froitzheim, S. Merbold, R. Ostilla-Mónico, and C. Egbers. "Angular momentum transport and flow organization in Taylor-Couette flow at radius ratio of $\eta = 0.357$". In: *Phys. Rev. Fluids* 4 (2019), p. 084605.

[42] A. Froitzheim, R. Ezeta, S. Merbold, C. Egbers, D. Lohse, and C. Sun. "Formation of large scale vortices in the torque maximum region in Taylor-Couette flow." In: *Fachtagung Experimentelle Strömungsmechanik, Deutsche Gesellschaft für Laser-Anemometrie GALA e.V.* 23. Cottbus, 2017b.

[43] A. Froitzheim, R. Ezeta, S. G. Huisman, S. Merbold, C. Sun, D. Lohse, and C. Egbers. "Statistics, plumes and azimuthally travelling waves in ultimate Taylor–Couette turbulent vortices". In: *J. Fluid Mech.* 876 (2019), 733–765.

[44] A. Froitzheim, S. Merbold, and C. Egbers. "Taylor-Couette turbulence at wide and very wide gaps." In: *Fachtagung Experimentelle Strömungsmechanik, Deutsche Gesellschaft für Laser-Anemometrie GALA e.V.* 24. Karlsruhe, 2016.

[45] A. Froitzheim, S. Merbold, and C. Egbers. "Velocity profiles, flow structures and scalings in a wide-gap turbulent Taylor-Couette flow." In: *J. Fluid Mech.* 831 (2017a), pp. 330–357.

[46] W. Gander, G. H. Golub, and R. Strebel. "Least-square fitting of circles and ellipses." In: *BIT* 43 (1994), pp. 558–578.

[47] G. Gauthier, P. Gondret, and M. Rabaud. "Motions of anisotropic particles: Application to visualization of three-dimensional flows." In: *Phys. Fluids* 10 (1998), 2147–2154.

[48] A. Goharzadeh and I. Mutabazi. "Experimental characterization of intermittency regimes in the Couette-Taylor system." In: *Eur. Phys. J. B* 19 (2001), pp. 157–162.

[49] S. Grossmann and D. Lohse. "Multiple scaling in the ultimate regime of thermal convection." In: *Phys. Fluids* 23 (2011), p. 045108.

[50] S. Grossmann and D. Lohse. "Scaling in thermal convection: a unifying theory." In: *J. Fluid Mech.* 407 (2000), pp. 27–56.

[51] S. Grossmann, D. Lohse, and C. Sun. "High-Reynolds Number Taylor-Couette Turbulence." In: *Annu. Rev. Fluid Mech.* 48 (2016), pp. 53–80.

[52] S. Grossmann, D. Lohse, and C. Sun. "Velocity profiles in strongly turbulent Taylor-Couette flow." In: *Phys. Fluids* 26 (2014), p. 025114.

[53] U. Harlander, T. von Larcher, Y. Wang, and C. Egbers. "PIV- and LDV- measurements of baroclinic wave interactions in a thermally driven rotating annulus." In: *Exp. Fluids* 51 (2011), pp. 37–49.

[54] B. Hof, J. Westerweel, T. M. Schneider, and B. Eckhardt. "Finite lifetime of turbulence in pipe flow." In: *Nature* 443 (2006), pp. 59–62.

[55] S. G. Huisman, D. P. M. van Gils, and C. Sun. "Applying laser Doppler anemometry inside a Taylor-Couette geometry using a ray-tracer to correct for curvature effects." In: *Eur. J. Mech. -B/Fluids* 36 (2012b), pp. 115–119.

[56] S. G. Huisman, S. Scharnowski, C. Cierpka, C. J. Kaehler, D. Lohse, and C. Sun. "Logarithmic Boundary Layers in Strong Taylor-Couette Turbulence." In: *Phys. Rev. Lett.* 110 (2013b), p. 264501.

[57] S. G. Huisman, R. C. A. van der Veen, C. Sun, and D. Lohse. "Multiple states in highly turbulent Taylor-Couette flow." In: *Nat.Commun.* 5 (2014), p. 3820.

[58] S. G. Huisman, D. Lohse, and C. Sun. "Statistics of turbulent fluctuations in counter-rotating Taylor-Couette flows." In: *Phys. Rev. E* 88 (2013a), p. 063001.

[59] S. G. Huisman, R. C. A. van der Veen, G.-W. H. Bruggert, D. Lohse, and C. Sun. "The boiling Twente Taylor-Couette (BTTC) facility: Temperature controlled turbulent flow between independently rotating, coaxial cylinders." In: *Rev. Sci. Instrum.* 86 (2015), p. 065108.

[60] S. G. Huisman, D. P. M. van Gils, S. Grossmann, C. Sun, and D. Lohse. "Ultimate turbulent Taylor-Couette flow." In: *Phys. Rev. Lett.* 108 (2012), p. 024501.

[61] J. Jimenez. "Cascades in wall-bounded turbulence". In: *Ann. Rev. Fluid. Mech.* 44 (2012), pp. 27–45.

[62] R. D. Keane and R. J. Adriane. "Optimization of particle image velocimeters. Part I: Double pulsed systems." In: *Meas. Sci. Technol.* 1 (1990), pp. 1202–1215.

[63] R. D. Keane and R. J. Adriane. "Optimization of particle image velocimeters. Part II: Multiple pulsed systems." In: *Meas. Sci. Technol.* 2 (1991), pp. 963–974.

[64] R. D. Keane and R. J. Adriane. "Theory of cross-correlation analysis of PIV images." In: *Appl. Sci. Res.* 49 (1992), pp. 191–215.

[65] G. P. King, Y. Li, W. Lee, and H. L. Swinney. "Wave speeds in wavy Taylor-vortex flow." In: *J. Fluid. Mech.* 141 (1984), pp. 365–390.

[66] D. P. Lathrop, J. Fineberg, and H. S. Swinney. "Transition to shear-driven turbulence in Couette-Taylor flow." In: *Phys. Rev. A* 46 (1992), pp. 6390–405.

[67] C. Leclercq, B. Pier, and J. F. Scott. "Temporal stability of eccentric Taylor-Couette-Poiseuille flow." In: *J. Fluid Mech.* 733 (2013), pp. 68–99.

[68] G. S. Lewis and H. L. Swinney. "Velocity structure functions, scaling and transitions in high-Reynolds-number Couette-Taylor flow." In: *Phys. Rev. E* 59 (1999), pp. 5457–67.

[69] W. V. R. Malkus and G. Veronis. "Finite amplitude cellular convection." In: *J. Fluid. Mech.* 4 (1958), pp. 225–260.

[70] A. Mallock. "Determination of the viscosity of water." In: *Proc. R. Soc. Lond.* 45 (1888), pp. 126–132.

[71] A. Mallock. "Experiments on Fluid Viscosity." In: *Phil. Trans. Roy. Soc. Lond.* A187 (1896), pp. 41–56.

[72] P. S. Marcus. "Simulation of Taylor-Couette fow. Part 2. Numerical results for wavy-vortex flow with one travelling wave." In: *J. Fluid. Mech.* 146 (1984), pp. 65–113.

[73] S. L. Marple. "Computing the discrete-time analytic signal via FFT." In: *IEEE Trans. Sig. Proc.* 47(9) (1999), pp. 2600–2603.

[74] B. Martínez-Arias, J. Peixinho, O. Crumeyrolle, and I. Mutabazi. "Effect of the number of vortices on the torque scaling in Taylor-Couette flow." In: *J. Fluid Mech.* 748 (2014), pp. 756–767.

[75] Ivan Marusic and Jason P. Monty. "Attached Eddy Model of Wall Turbulence". In: *Annual Review of Fluid Mechanics* 51.1 (2019), null.

[76] S. Merbold, A. Froitzheim, and Ch. Egbers. "Flow pattern and angular motion transport in a wide gap Taylor-Couette flow." In: *Schriftenreihe aus dem Institut für Strömungsmechanik*. Strömungstechnische Tagung 10. Dresden, 2014.

[77] S. Merbold, A. Froitzheim, and C. Egbers. "Torque in turbulent Taylor-Couette flow for small radius ratios". In: *Proceedings of the 5th International Conference on Experimental Fluid Mechanics*. Ed. by C. J. Kähler, R. Hain, S. Scharnowski, and T. Fuchs. International Conference on Experimental Fluid Mechanics. Munich, 2018, pp. 598–602.

[78] S. Merbold, H. J. Brauckmann, and C. Egbers. "Torque measurements and numerical determination in differentially rotating wide gap Taylor-Couette flow." In: *Phys. Rev. E* 87 (2013), p. 023014.

[79] W. Merzkirch. *Flow Visualization*. 2nd ed. Orlando, Florida: Academic Press, Inc., 1987.

[80] W. Nitsche and A. Brunn. *Strömungsmesstechnik*. 2nd ed. Berlin Heidelberg: Springer, 2006.

[81] H. Nobach and E. Bodenschatz. "Limitations of accuracy in PIV due to individual variations of particle image intensities." In: *Exp. Fluids* 47 (2009), pp. 27–38.

[82] R. Ostilla, R. J. A. M. Stevens, S. Grossmann, R. Verzicco, and D. Lohse. "Optimal Taylor-Couette flow: direct numerical simulations." In: *J. Fluid Mech.* 719 (2013), pp. 14–46.

[83] R. Ostilla-Mónico, E. P. van der Poel, R. Verzicco, S. Grossmann, and D. Lohse. "Boundary layer dynamics at the transition between the classical and ultimate regime of Taylor-Couette flow." In: *Phys. Fluids* 26 (2014), p. 015114.

[84] R. Ostilla-Mónico, D. Lohse, and R. Verzicco. "Effect of roll number on the statistics of turbulent Taylor-Couette flow". In: *Physical review fluids* 1.5 (2016), p. 054402.

[85] R. Ostilla-Mónico, R. Verzicco, and D. Lohse. "Effects of the computational domain size on direct numerical simulations of Taylor-Couette turbulence with stationary outer cylinder". In: *Physics of Fluids* 27.2 (2015), p. 025110.

[86] R. Ostilla-Mónico, E. P. van der Poel, R. Verzicco, S. Grossmann, and D. Lohse. "Exploring the phase diagram of fully turbulent Taylor-Couette flow." In: *J. Fluid Mech.* 761 (2014), pp. 1–26.

[87] R. Ostilla-Mónico, S. G. Huisman, T. J. G. Janninik, D. P. M. van Gils, R. Verzicco, S. Grossmann, and D. Lohse. "Optimal Taylor-Couette flow: radius ratio dependence." In: *J. Fluid Mech.* 747 (2014), pp. 1–29.

[88] R. Ostilla-Mónico, R. Verzicco, S. Grossmann, and D. Lohse. "The near-wall region of highly turbulent Taylor-Couette flow." In: *J. Fluid Mech.* 788 (2016), pp. 95–117.

[89] A. Pandey, J. D. Scheel, and J. Schumacher. "Turbulent superstructures in Rayleigh-Bénard convection." In: *Nat. Commun.* 9 (2018), p. 2118.

[90] M. S. Paoletti and D. P. Lathrop. "Angular momentum transport in turbulent flow between independently rotating cylinders." In: *Phys. Rev. Lett.* 106 (2011), p. 024501.

[91] M. S. Paoletti, D. P. M. van Gils, S. Grossmann, C. Sun, and D. Lohse. "Angular momentum transport and turbulence in laboratory models of Keplerian flows." In: *Astron. Astrophys.* 547 (2012), A649.

[92] R. L. Pfeffer, J. Ahlquist, R. Kung, Y. Chang, and G. Li. "A study of baroclinic wave behaviour over bottom topography using complex principal component analysis of experimental data." In: *Atmos. Sci.* 47 (1990), pp. 67–81.

[93] S. B. Pope. *Turbulent flows.* Cambridge University Press, 2000.

[94] L. Prandtl. "Entstehung von Wirbeln bei Wasserströmungen - 1. Entstehung von Wirbeln und künstliche Beeinflussung der Wirbelbildung." In: *Institut für Wissenschaftlichen Film (IWF), Göttingen, C1* (1936).

[95] L. Prandtl. "Über Flüssigkeitsbewegung bei sehr kleiner Reibung." In: *Verhandlungen des III. Int. Math. -Kongresses.* Leipzig: Teubner, 1905, pp. 484–491.

[96] W. H. Press, S. A. Teukolsky, W. T. Vetterling, and B. P. Flannery. *Numerical Recipes: The Art of Scientific Computing.* 3rd ed. Cambridge University Press, 2007, p. 653. ISBN: 978-0-521-88068-8.

[97] A. Prigent, B. Dubrulle, O. Dauchot, and I. Mutabazi. "The Taylor-Couette Flow: The Hydrodynamic Twin of Rayleigh-Bénard Convection. Springer Tracts in Modern Physics". In: *Dynamics of Spatio-Temporal Cellular Structures.* Ed. by I. Mutabazi, J.E. Wesfreid, and E. Guyon. Vol. 207. Springer, New York, NY, 2006.

[98] I. Procaccia, E. S. C. Ching, P. Constantin, L. P. Kadanoff, A. Libchaber, and X. Z. Wu. "Transition to convective turbulence: The role of thermal plumes". In: *Phys. Rev. A* 44 (1991), pp. 8091–8102.

[99] F. Ravelet, R. Delfos, and J. Westerweel. "Influence of global rotation and Reynolds number on the large-scale features of a turbulent Taylor-Couette flow." In: *Phys. Fluids* 22 (2010), p. 055103.

[100] H. Reichenbach. "Contributions of Ernst Mach to fluid mechanics." In: *Ann. Rev. Fluid Mech.* 15 (1983), pp. 1–28.

[101] O. Reynolds. "An experimental investigation of the circumstances which determine whether the motion of water shall be direct or sinuous, and of the law of resistance in parallel channels." In: *Proc. R. Soc. Lond.* 35 (1883), pp. 84–99.

[102] J. C. Rotta. *Turbulente Strömungen - Eine Einführung in die Theorie und ihre Anwendungen.* Unv.-Verl. Göttingen, 1972.

[103] G. Rüdiger, M. Schultz, F. Stefani, and R. Hollerbach. "Magnetorotational instability in Taylor-Couette flows between cylinders with finite electrical conductivity". In: *Geophysical & Astrophysical Fluid Dynamics* 112.4 (2018), pp. 301–320.

[104] F. Scarano. "Tomographic PIV: principles and practice." In: *Meas. Sci. Technol.* 24 (2013), p. 012001.

[105] E. Schartman, H. Ji, and M. J. Burin. "Development of a Couette-Taylor flow device with active minimization of secondary circulation." In: *Rev. Sci. Instrum.* 80 (2009), p. 024501.

[106] E. Schartmann, H. Ji, M. J. Burin, and J. Goodman. "Stability of quasi-Keplerian shear flow in a laboratory experiment." In: *Astron. Astrophys.* 543 (2012), A94.

[107] H. Schlichting and K. Gersten. *Grenzschicht-Theorie.* 10th ed. Berlin Heidelberg: Sprin-ger, 2006. ISBN: 3-54023004-1.

[108] S. Seghir-Qualil, S. Harmand, and D. Laloy. "Study of the thermal behavior of a synchronous motor with permanent magnets." In: *Int. J. Eng.* 3 (1990), pp. 229–256.

[109] X.-D. Shang, X.-L. Qiu, P. Tong, and K.-Q. Xia. "Measurements of the local convective heat flux in turbulent Rayleigh-Bénard convection." In: *Phys. Rev.* 70 (2004), p. 026308.

[110] O. Shishkina and C. Wagner. "Local heat fluxes in turbulent Rayleigh-Bénard convection." In: *Phys. Fluids* 19 (2007), p. 085107.

[111] G. P. Smith and A. A. Townsend. "Turbulent Couette flow between concentric cylinders at large Taylor numbers." In: *J. Fluid. Mech.* 123 (1982), pp. 187–217.

[112] Alexander J. Smits, Beverley J. McKeon, and Ivan Marusic. "High-Reynolds Number Wall Turbulence". In: *Annu. Rev. Fluid Mech.* 43 (2011), pp. 353–375.

[113] J. H. Spurk and N. Aksel. *Fluid Mechanics.* 2nd ed. Berlin Heidelberg: Springer, 2008, p. 5. ISBN: 978-3-540-73536-6.

[114] Richard J. A. M. Stevens, Alexander Blass, Xiaojue Zhu, Roberto Verzicco, and Detlef Lohse. "Turbulent thermal superstructures in Rayleigh-Bénard convection". In: *Phys. Rev. Fluids* 3 (4 2018), p. 041501.

[115] R. D. Suanto, Q. Zheng, and X.-H. Yan. "Complex singular value decomposition analysis of equatorial waves in the Pacific observed by TOPEX/Poseidon altimeter." In: *J. Atmos. Oceanic Technol.* 15 (1998), pp. 764–774.

[116] H. L. Swinney and J. P. Gollub. *Hydrodynamic Instabilities and the Transition to Turbulence.* 2nd ed. Springer, 1985.

[117] K. Taira, S. L. Brunton, S. T. M. Dawson, C. W. Rowley, T. Colonius, B. J. McKeon, O. T. Schmidt, S. Gordeyev, V. Theofilis, and L. S. Ukeiley. "Modal Analysis of Fluid Flows: An Overview." In: *AIAA* 55(12) (2017), pp. 4013–4041.

[118]   Y. Takeda. "Quasi-periodic state and transition to turbulence in a rotating Couette system." In: *J. Fluid. Mech.* 389 (1999), pp. 81–99.

[119]   T. V. Tamhane, J. B. Joshi, U. K. Mudali, R. Natarajan, and R. N. Patil. "Axial mixing in annular centrifugal extractors." In: *Chem. Eng. J.* 207-208 (2012), pp. 462–472.

[120]   G. I. Taylor. "Fluid Friction between Rotating Cylinders. I. Torque Measurements." In: *Proc. R. Soc. Lond.* A157 (1936), pp. 546–564.

[121]   G. I. Taylor. "Stability of a Viscous Liquid Contained between Two Rotating Cylinder." In: *Phil. Trans. Roy. Soc. London* 223 (1923), pp. 289–343.

[122]   S. T. Thoroddsen and C. W. van Atta. "Exponential tails and skewness of density-gradient PDF's in stable stratified turbulence." In: *J. Fluid Mech.* 244 (1992), pp. 547–566.

[123]   S. Tokgoz, G. E. Elsinga, R. Delfos, and J. Westerweel. "Experimental investigation of torque scaling and coherent structures in turbulent Taylor-Couette flow." In: *J. Phys.: Conf. Ser* 318 (2011), p. 082018.

[124]   S. Tokgoz, G. E. Elsinga, R. Delfos, and J. Westerweel. "Spatial resolution and dissipation rate estimation in Taylor-Couette flow for tomographic PIV." In: *Exp. Fluids* 53 (2012), pp. 561–583.

[125]   C. Tropea, L. Alexander, and L. Yarin. *Handbook of Experimental Fluid Mechanics.* Berlin Heidelberg: Springer, 2007. ISBN: 978-3-540-25141-5.

[126]   E. P. van der Poel, R. Ostilla-Monico, J. Donners, and R. Verzicco. "A pencil distributed finite difference code for strongly turbulent wall-bounded flows". In: *Comp. Fluids* 116 (2015), pp. 10–16.

[127]   R. C. A. van der Veen, S. G. Huisman, O.-Y-Dung, H. L. Tang, C. Sun, and D. Lohse. "Exploring the phase space of multiple states in highly turbulent Taylor-Couette flow." In: *Phys. Rev. Fluids* 1 (2016a), p. 024401.

[128]   R. C. A. van der Veen, S. G. Huisman, S. Merbold, U. Harlander, C. Egbers, D. Lohse, and C. Sun. "Taylor-Couette turbulence at radius ratio 0.5: scaling, flow structures and plumes." In: *J. Fluid Mech.* 799 (2016b), pp. 334–351.

[129]   M. van Dyke. *An Album of Fluid Motion.* Stanford, California: The Parabolic Press., 1982.

[130]   D. P. M. van Gils, S. G. Huisman, S. Grossmann, C. Sun, and D. Lohse. "Optimal Taylor-Couette turbulence." In: *J. Fluid. Mech.* 706 (2012), pp. 118–149.

[131]   D. P. M. van Gils, G.-W. Bruggert, D. P. Lathrop, C. Sun, and D. Lohse. "The Twente turbulent Taylor-Couette (T3C) facility: Strongly turbulent (multiphase) flow between two independently rotating cylinders." In: *Rev. Sci. Instrument* 82 (2011), p. 025105.

[132]   D. P. M. van Gils, S. G. Huisman, G.-W. Bruggert, C. Sun, and D. Lohse. "Torque scaling in turbulent Taylor-Couette flow with co- and counterrotating cylinders." In: *Phys. Rev. Lett.* 106 (2011), p. 024502.

[133]   R. van Hout and J. Katz. "Measurements of mean flow and turbulence characteristics in high-Reynolds number counter-rotating Taylor-Couette flow." In: *Phys. Fluids* 23 (2011), p. 105102.

[134] R. Verzicco and P. Orlandi. "A finite-difference scheme for three-dimensional incompressible flow in cylindrical coordinates". In: *J. Comput. Physics* 123 (1996), pp. 402–413.

[135] R. W. Walden and R. J. Donelly. "Reemergent Order of Chaotic Circular Couette Flow ." In: *Phys. Rev. Lett.* 42 (1979), pp. 301–304.

[136] L. Wang, M. G. Olsen, R. Dennis, and R. D. Vigil. "Reappearance of azimuthal waves in turbulent Taylor-Couette flow at large aspect ratio." In: *Chem. Eng. Sci.* 60 (2005), p. 5555.

[137] P. D. Weidman. "Measurement techniques in laboratory rotating flows." In: *Advances in Fluid Mechanics Measurements*. Ed. by M. Gad el Hak. Vol. 45. Springer Science & Business Media, 2013.

[138] F. Wendt. "Turbulente Strömungen zwischen zwei rotierenden konaxialen Zylindern." In: *Ing. Arch.* 4 (1933), p. 577.

[139] S. T. Wereley and R. M. Lueptow. "Spatio-temporal character of non-wavy and wavy Taylor-Couette flow." In: *J. Fluid Mech.* 364 (1998), pp. 59–80.

[140] S. T. Wereley and R. M. Lueptow. "Velocity field for Taylor-Couette flow with an axial flow." In: *Phys. Fluids* 11(12) (1999), pp. 3637–3649.

[141] J. Westerweel, G. E. Elsinga, and R. J. Adrian. "Particle Imgage Velocimetry for Complex and Turbulent Flows." In: *Annu. Rev.. Fluid Mech.* 45 (2013), pp. 409–436.

[142] E. M. Withjack and C. Chen. "An experimental study of Couette instability of stratified fluids." In: *J. Fluid Mech.* 66 (1974), pp. 725–737.

[143] V. Yakhot. "Probability distributions in high-Rayleigh number Bénard convection." In: *Phys. Rev. Lett.* 63 (1989), pp. 1965–1967.

Acknowledgements

This thesis was created during my time as scientific staff member at the Department of Aerodynamics and Fluid Mechanics at the Brandenburg University of Technology Cottbus-Senftenberg and was funded by the German Research Foundation (DFG FOR 1182, grant no. EG100/15-2) in the framework of the project: "Wandnahe Transport- und Strukturbildungsprozesse in turbulenten Rayleigh-Bénard-, Taylor-Couette- und Rohrströmungen". In addition, a research stay at the University of Twente was funded by the European High-Performance Infrastructures in Turbulence (EuHIT) consortium for the project: "Formation of large-scale vortices in the torque maximum region in Taylor-Couette flow".

First of all, I would like to thank my supervisor Prof. Christoph Egbers for his continuous support and guidance. He encouraged me to find my own way through the field of fluid mechanics and always had an open door when I needed advice. During my doctoral time at your institute, I could develop not only scientifically but also personally.

Further, I would like to thank Prof. Detlef Lohse for his willingness to become the second referee of my thesis and the successful collaboration with his Physics of Fluids group within the EuHIT-framework. Especially, I am very thankful for the hospitality and the stimulating scientific discussions during my research stay in Twente.

I am also very grateful to Rodolfo Ostilla-Mónico, who provided me numerical data for validation and more detailed interpretation of my own measurements. I really appreciated his fruitful ideas and contributions to this thesis. In addition, I want to thank Rodrigo Ezeta Aparicio, Sander G. Huisman and Chao Sun for the exciting time in the lab and the valuable meetings in Twente and the continuing communication afterwards.

I want to thank all my colleagues in the Department of Aerodynamics and Fluid Mechanics for the pleasant working environment, the manifold help with the everyday problems of an experimenter, unforgettable conference travels and much more: Uwe, Florian Z., Martin, Michael, Torsten, Andreas K., Andreas S., Vasyl, El-Sayed, Peter, Gabriel, Marcel, Zeinab, Mohamed, Markus, Antoine, Costanza, Wenchao, Robin, Ludwig Florian P., Vadim, I really had a great time with you. Furthermore, I want to thank Silke Kaschwich for her constant help with all kinds of bureaucracy and Heinz-Jörg Wengler for keeping my experiment running for four and a half years. In particular, I would like to thank my group leader, TC-mate and friend Sebastian Merbold. It was a really pleasure to work with you together, to learn from you about so many things and to have someone who shares my enthusiasm for vortices behind acrylic cylinders. Beside, I would like to thank my office mates and friends Amir Shahirpour, Gazi Hasanuzzaman and Stefan Richter, who have made many working days flying by. Also many thanks to Eduard Pascu and Ngoc Thanh Trung Pham for the great support during different measurement campaigns.

I especially want to thank my parents Christine and Michael, who supported me unconditionally all my life and whom I owe so much. Finally, I can not put into words how grateful I am to my wife Stefanie. She was there for me in good and bad times, supported and encouraged me and provided me distraction whenever I needed it. Above all, she gave me our son Felix, whose cheerful and lovely nature motivates me every single day.

List of Publications

Journal articles

- A. Froitzheim, S. Merbold, R. Ostilla-Mónico and C. Egbers, *Angular momentum transport and flow organization in Taylor-Couette flow at radius ratio of $\eta = 0.357$*, Phys. Rev. Fluids 4 (2019), 084605 [41].

- A. Froitzheim, R. Ezeta, S.G. Huisman, S. Merbold, C. Sun, D. Lohse and C. Egbers, *Statistics, plumes and azimuthally traveling waves in ultimate Taylor-Couette turbulent vortices*, J. Fluid Mech. 876 (2019), 733-765 [43].

- A. Froitzheim, S. Merbold and C. Egbers, *Velocity profiles, flow structures and scalings in a wide-gap turbulent Taylor-Couette flow*, J. Fluid Mech. 831 (2017), 330-357 [45].

Conference articles and contributions

- F. Zaussinger, Ch. Egbers, T. Driebe, P. Heintzmann, P. Canfield, M. Meier, V. Travnikow, A. Froitzheim, P. Haun, *ATMOFLOW - Simulating atmosphere flows on the international space station. Part II: Experiments and numerical simulations*, Proceedings of the 69th International Astronautical Congress, October 1-5th 2018, Bremen, Germany.

- A. Froitzheim, S. Merbold, Ch. Egbers, R. Ezeta, D. Lohse, C. Sun, *Large scale vortices in the Taylor-Couette flow*, Proceedings of the 12th European Fluid Mechanics Conference, September 7-9th 2018, Vienna, Austria.

- C. Egbers, F. Zaussinger, P. Haun, P. Canfield, V. Travnikov, A. Froitzheim, *Convection in the spherical gap under micro-gravity conditions: From Earth's mantle to atmospheric flows*, Proceedings of the 20th International Couette-Taylor Workshop, July 11-13th 2018, Marseille, France.

- A. Froitzheim, S. Merbold, Ch. Egbers, R. Ezeta, D. Lohse, C. Sun, *Large scale vortices in the Taylor-Couette flow*, Proceedings of the 20th International Couette-Taylor Workshop, July 11-13th 2018, Marseille, France.

- S. Merbold, A. Froitzheim, C. Egbers, *Torque in turbulent Taylor-Couette flow for small radius ratios*, Proceedings of the 5th International Conference on Experimental Fluid Mechanics, July 2-4th 2018, Munich, Germany, pp. 598-602, ISBN 978-3-943207-32-3.

- A. Froitzheim, R. Ezeta, S. Merbold, Ch. Egbers, D. Lohse, C. Sun, *Formation of large scale vortices in the torque maximum region in Taylor-Couette flow*, Proceedings of the 25th GALA Fachtagung Experimentelle Strömungsmechanik, September 5-7th 2017, Karlsruhe-Ettlingen, Germany, ISBN 978-3-9816764-3-3.

- A. Froitzheim, S. Merbold, Ch. Egbers, *Torque measurements in a Taylor-Couette experiment with very small radius ration of 0.357*, Proceedings of the 16th European Turbulence Conference, August 21-24th 2017, Stockholm, Sweden.

140

- A. Froitzheim, S. Merbold, Ch. Egbers, *Flow organization in the region of the torque maximum in a wide gap Tayor-Couette experiment*, Proceedings of the 88th Annual Meeting of the International Association of Applied Mathematics and Mechanics GAMM, March 6-10th 2017, Weimar, Germany.

- A. Froitzheim, S. Merbold, Ch. Egbers, *Momentum transport and flow structures in a wide gap fully turbulent Taylor-Couette flow*, Proceedings of the 11th European Fluid Mechanics Conference, September 13-16th 2016, Sevilla, Spain.

- A. Froitzheim, S. Merbold, Ch. Egbers, *Taylor-Couette turbulenz at wide and very wide gap*, Proceedings of the 24th GALA Fachtagung Experimentelle Strömungsmechanik, September 6-8th 2016, Cottbus, Germany, ISBN 978-3-9816764-2-6.

- A. Froitzheim, S. Merbold, Ch. Egbers, *Experimental Investigations of Taylor-Couette flow at BTU Cottbus*. Proceedings of the School Instabilities and Turbulence in Strato-Rotational Flows, July 11-13th 2016, Le Havre, France.

- A. Froitzheim, S. Merbold, Ch. Egbers, *PIV Messungen in einer Taylor-Couette Strömung mit weitem Spalt*, Proceedings of the 23th GALA Fachtagung Lasermethoden in der Strömungsmesstechnik, September 8-10th 2015, Dresden, Germany, ISBN 978-3-9816764-1-9.

- A. Froitzheim, S. Merbold, Ch. Egbers, *Torque measurements and flow visualisation in a wide gap Taylor-Couette Flow*, Proceedings of the 15th European Turbulence Conference, August 25-28th 2015, Delft, The Netherlands.

- B. Martinez-Arias, J. Peixinho, A. Froitzheim, S. Merbold, Ch. Egbers, I. Mutabazi, *Influence of the radius ratio on the torque in turbulent Taylor-Couette flow*, Proceedings of the 19th International Couette-Taylor Workshop, June 22-24th 2015, Cottbus, Germany.

- S. Merbold, A. Froitzheim, Ch. Egbers, *Turbulent Taylor-Couette flow for very wide gaps*, Proceedings of the 19th International Couette-Taylor Workshop, June 22-24th 2015, Cottbus, Germany.

- A. Froitzheim, S. Merbold, Ch. Egbers, *Wide gap Taylor-Couette flow*, Proceedings of the 19th International Couette-Taylor Workshop, June 22-24th 2015, Cottbus, Germany.

- S. Merbold, A. Froitzheim, Ch. Egbers, *Flow Pattern and angular motion transport in a wide gap Taylor-Couette flow*, 10. Band aus der Schriftenreihe aus dem Institut für Strömungsmechanik der Strömungstechnische Tagung, Dresden, Germany, ISBN 978-3-944331-78-2.

List of attended Conferences

- 12th European Fluid Mechanics Conference, September 7-9th 2018, Vienna, Austria.

- 20th International Couette-Taylor Workshop, July 11-13th 2018, Marseille, France.

- 25th GALA Fachtagung Experimentelle Strömungsmechanik, September 5-7th 2017, Karlsruhe-Ettlingen, Germany.

- 16th European Turbulence Conference, August 21-24th 2017, Stockholm, Sweden.

- 88th Annual Meeting of the International Association of Applied Mathematics and Mechanics GAMM, March 6-10th 2017, Weimar, Germany.

- 11th European Fluid Mechanics Conference, September 13-16th 2016, Sevilla, Spain.

- 24th GALA Fachtagung Experimentelle Strömungsmechanik, September 6-8th 2016, Cottbus, Germany.

- EuHIT Turbulence Workshop, May 2-4th 2016, Göttingen, Germany.

- 23th GALA Fachtagung Lasermethoden in der Strömungsmesstechnik, September 8-10th 2015, Dresden, Germany.

- 15th European Turbulence Conference, August 25-28th 2015, Delft, The Netherlands.

- 19th International Couette-Taylor Workshop, June 22-24th 2015, Cottbus, Germany.

Curriculum vitae

Andreas Froitzheim was born on March 17 in 1988 in Wiesbaden and grew up in Auringen. He graduated from school in 2007 and studied afterwards from 2008 to 2014 Mechanical and Process Engineering at the Technical University of Darmstadt. He obtained his Bachelor of Science degree in 2011 after his Bachelor-Thesis "Construction of nucleation sites for bubble coalescence by laser surface treatment and whose investigation in an existing test facility" at the Institute of Technical Thermodynamics. In the subsequent Master's program he focused on the topics aerodynamics and fluid mechanics, turbo-machinery as well as thermodynamics and power engineering. After completing his Master-Thesis with the title "Automated CAD-based optimization of high pressure turbine flow" at Rolls-Royce Germany in the Department of Turbine Aerothermal, he obtained the Master of Science degree in 2014. In the same year, he started his PhD at the Department of Aerodynamics and Fluid Mechanics of the Brandenburg University of Technology Cottbus-Senftenberg. His research topic was part of the project "Wandnahe Transport- und Strukturbildungsprozesse in turbulenten Rayleigh-Bénard-, Taylor-Couette- und Rohrströmungen" funded by the German Research Foundation and included the experimental investigation of the turbulent Taylor-Couette flow using visualization techniques, torque measurements and particle image velocimetry. In 2019, he completed his doctorate with the award of the academic degree of Doctor of Engineering.

Lebenslauf

Andreas Froitzheim wurde am 17. März 1988 in Wiesbaden geboren und wuchs in Auringen auf. Die allgemeine Hochschulreife erwarb er im Jahr 2007 und studierte danach von 2008 bis 2014 Maschinenbau an der Technischen Universität Darmstadt. Den Bachelorstudiengang schloss er 2011 mit seiner Bachelorarbeit „Herstellung von Keimstellen zur Blasenkoaleszenz mittels Laser-Oberflächenstrukturierung und deren Untersuchung an einer bestehenden Versuchsanlage"am Institut für Technische Thermodynamik als Bachelor of Science ab. Im sich anschließenden Masterstudiengang legte er seine Schwerpunkte auf die Themengebiete Strömungsmechanik und Aerodynamik, Turbomaschinen sowie Thermodynamik und Energiesysteme. Den Master of Science erlangte er im Jahr 2014 nach Beendigung seiner Masterarbeit bei Rolls-Royce Deutschland in der Abteilung Turbine Aerothermal mit dem Titel „Automatisierte CAD-basierte Optimierung der Strömung von Hochdruckturbinen". Noch im selben Jahr begann er seine Promotion am Lehrstuhl Aerodynamik und Strömungslehre der Brandenburgischen Technischen Universität Cottbus-Senftenberg. Sein Forschungsthema war Teil eines von der Deutschen Forschungsgesellschaft geförderten Projektes mit dem Namen „Wandnahe Transport- und Strukturbildungsprozesse in turbulenten Rayleigh-Bénard-, Taylor-Couette- und Rohrströmungen" und umfasste die experimentelle Untersuchung der turbulenten Taylor-Couette-Strömung mittels Visualisierungstechniken, Drehmomentmessungen sowie Particle Image Velocimetry. Im Jahr 2019 schloss er seine Promotion mit der Erlangung des akademischen Grades eines Doktors der Ingenieurswissenschaften ab.